FMCG PACKAGING
INNOVATIONS

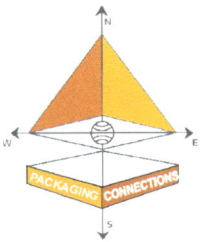

Sanex Packaging Connections Pvt. Ltd.
www.packagingconnections.com

Copyright

Published by :

Sanex Packaging Connections Pvt. Ltd.

An ISO 9001 : 2008 Certified Organisation

117, Suncity Tower, Sector-54

Golf Course Road, Gurgoan-122 002.

Tel : +91 124 4965770

Fax : + 91 124 41433951

e-mail : info@packagingconnections.com

Like us on Facebook : www.facebook.com/pconnection

ISBN : 9788192792040

FMCG INNOVATIONS

List of Contributors

Team www.PackagingConnections.com by Sanex Packaging Connections Pvt Ltd

Sandeep Kumar Goyal, Founder & CEO
Amita Venkatesh Valleesha, Associate: Scientific Affairs & Consultancy
Chhavi Goel, Associate: Research & Business Consulting
Bhaskar Ch, Technology Advisor e-business
Sonu Sheoran, Associate Research & Technology
Ashok Kumar, Programme Manager: KPO

FMCG INNOVATIONS

Table of Contents

Introduction by Sandeep Kumar Goyal	10
Design	**12-45**
ENEXY	14-15
BeeLoved	16-17
Stack your Wine	18-19
Diamond Bottle	20-21
Reseal it	22-23
Pouch up	24-25
3D Lenticulars	26-27
Smart Packaging	28-29
Choose your cereal	30-31
Easy Lift Lid	32-33
gogol mogol !	34-35
Instant soups in collapsible cups	36-37
Hot Sandwich in Flight	38-39
Rattle Tattle	40-41
On the go fresh Coffee	42-43
Reclosable and Tamper Evident Dispenser Carton	44-45
Break Through	**46-65**
Dosage Stick Packs with CR Features	48-49
On the Go salad!	50-51
Confectionery pack with magnetic closure	52-53
UV protective shrink sleeve	54-55
TE cap for Icecreams	56-57
Peelable Soup pouches	58-59
IML with barrier layer	60-61
New layout for fluted cartons	62-63
Sustainable Cupholder for Yoghurt and Museli	64-65
Technology	**66-77**
3D IML	68-69
Bombay Sapphire	70-71
Soft drink in Aerosol	72-73
Nor Cell (written in next sheet)	74-75
QR codes	76-77
Value Engineering	**78-93**
Die Cut Lid 2020	80-81
Seallite Lid (As written in the next sheet)	82-83
Soups in Tetra Recart	84-85
R - Laminates	86-87
Sustainable Square Tubes	88-89
Bread Packaging	90-91
Light Weight Bottle	92-93
Green Packaging	**94-103**
Light Cap	96-97
Post Consumer Recycled Ink	98-99
Eco friendly Cartons	100-101
Bamboo replaces EPE cushioning	102-103
Self Ready	**104-119**
On shelf cognac carton	106-107
Tidy Stacked Ice Cream Bars	108-109
Adjustable Partition for SRPs	110-111
3D Coffee Beans !	112-113
Profile Carton for Tea Pods	114-115
We believe what we see	116-117
Funstigator	118-119
Concept	**120-138**
Convenient Packaging	122-123
GIGS.2.GO	124-125
Draw the last tissue	126-127
Love Toothpaste tubes	128-129
Collapsible Instant Noodles Packaging	130-131
Break It Fast!	132-133
Take-away Stackable Packaging	134-35
Bloom Chips	136-137

FMCG INNOVATIONS

Introduction

This publication comes after the success of Ideas & Opportunities 2013 held on 19th July 2013 in India. Various innovations were presented during the one day workshop by the expert consulting team of Sanex packaging Connections Pvt Ltd popularly known as Team PackagingConnections.

Idea behind this is to bring the innovations to wider group of professionals to meet the mission of packaging knowledge sharing and that too cost effectively. We feel that this publication will further fill the project pipelines of companies and improve the standards of packaging. Many professionals either do not have the access or time to go through so many innovations together. So we think this publication will fill that gap. For your feedback please email directly to info@packagingconnections. com

With this, Enjoy Wonders Of Packaging!

Sandeep Kumar Goyal
Founder & CEO ,

DESIGN

Manufacturer/Designer

Dr. Alexander von Niessen
Chocal Aluminiumverpackungen
GmbH
Perlenweg 6D-73525 Schwäbisch
Gmünd
Postfach 1520, D-73505 Schwäbisch
Gmünd
Phone: +49 (0) 7171 / 1009-18
Fax: +49 (0) 7171 / 1009-8
Email: info@chocal.de

- Especially designed as per product contour
- The stunning pack shape and enhanced graphics are accomplished through clever design and use of an Alu/PE solvent free laminate.
- Tap into the strong demand for 'on the go' energy products.
- Important to the design is the tab opening which runs along the front of the bar rather than around it.
- The proximity to the can design is chosen deliberately

Manufacturer/Designer

Tamara Mihajlovic
Belgrade, Serbia
BeeLoved

- Such an exclusive product in many ways different and stand out from other ordinary honey packaging on market.

- Simple, clean, yet effective packaging design has been achieved by refracted form that draws its inspiration from nature.

- The structure resembles a piece of rock rolled away (detached piece of rock) , raw gemstone, diamond and hexagon shape - the honeycomb.

- Form draws attention in the retail space. In addition to the primary purpose, BEEloved honey can be used as a decoration and part of the interior.

- Beside proven quality and functional use, this is a product with equally decorative purpose.

- BEEloved honey contains in itself a piece of honeycomb, in the form of a pyramid.

- Honeycomb is not only Beneficial to

 **ZIP**

 **SNAP**

 **SIP**

Manufacturer/Designer

Stack wines
7148701227;
Matt Zimmer / Jodi wynn
info@drinkstack.com;
California

- Revolutionary product that vertically stacks four shatter-proof, individually packaged stemless PET wine glasses

- This forms the equivalent of one full bottle of wine.

- Concept is a complete deviation from any traditional wine packaging vehicle.

- These pre-filled "glasses" come, zipped into a convenient PET shrink-sleeve.

- Users simply: zip loose the sleeve, snap apart the glasses, and peel back a foil seal to sip.

- The new packaging utilizes language and graphics to show users intuitively how to use the product

- proclaiming, "zip, snap, sip" and the tagline "Take it with you."

- A see-through sleeve, pairing notes, and lifestyle suggestions for scenarios where Stack prevails over traditional

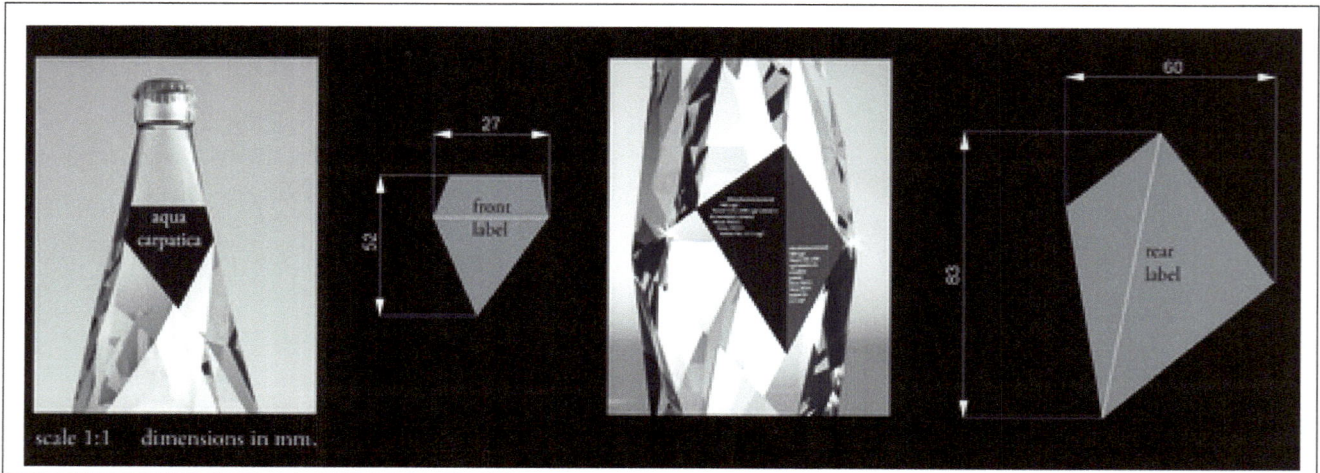

scale 1:1   dimensions in mm.

front label

rear label

27

52

60

83

Manufacturer/Designer

Cristiano Giuggioli
CL3V3R - Design Innovation
Milan, Italy
www.cl3v3r.it

True to its name "Diamond, this sparkling bottle gives the perfect impression of the purity.

The label is directly printed on the glass and the cap is a single use crown cap.

This also makes the recycling easy as the cap is already separated from the bottle

The aesthetic appeal of this bottle makes it reusable for other purpose after primary use.

Reseal it

Walkers
Pure Butter
MINI
SHORTBREAD
FINGERS

Strawberries
HEREFORDSHIRE, UK

Manufacturer/Designer

macfarlanegroup
21 Newton Place, Glasgow, G3 7PY
info@macfarlanegroup.com
Telephone: 0141 333 9666
Fax: 0141 333 1988

Reseal-it® is an innovative, patented label system that enables consumer packs to be easily opened and closed.

Re-Seal It® labels can be made in different sizes and shapes to fit most package designs.

The package also performs exceptionally well in vacuum and gas flushed food applications

For the consumer it presents packaging that is easier to open & access and easier re-close to retain the original quality of product.

Tamper evident features are available and there are a variety of label sizes and shapes

Pre-Applied web and Reseal it systems

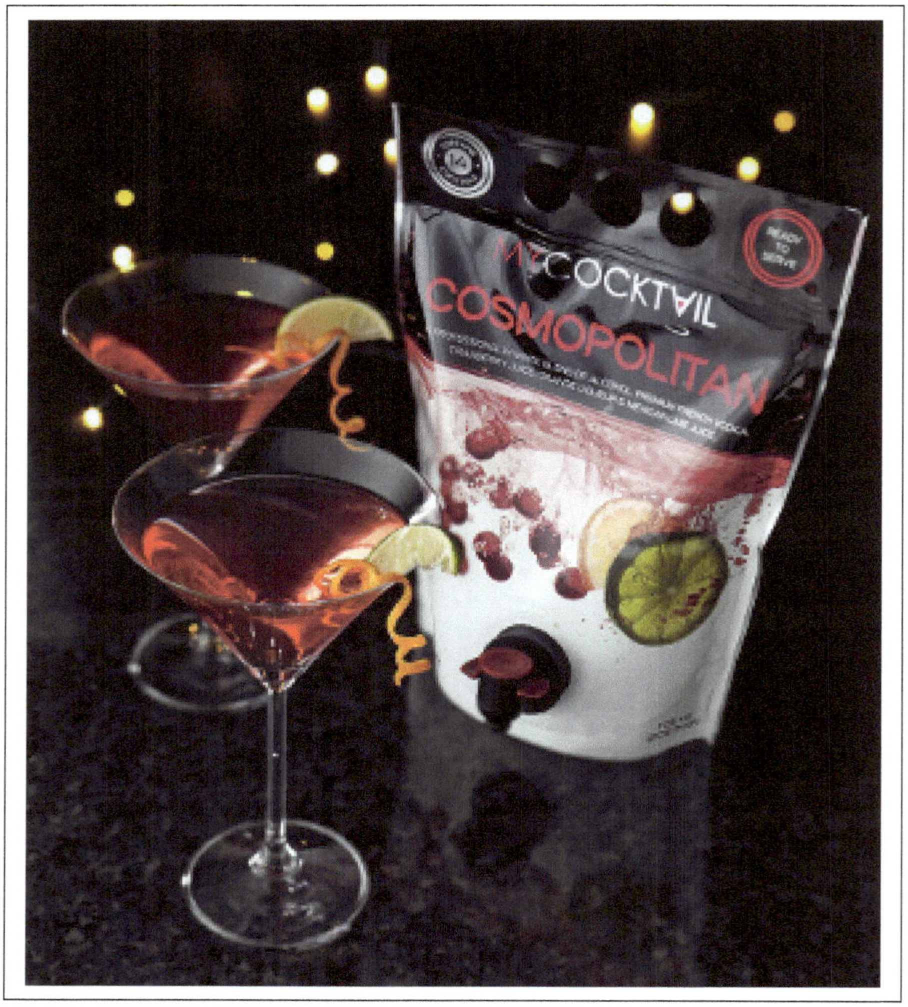
MYCOCKTAIL
COSMOPOLITAN
READY TO SERVE

Manufacturer/Designer

Smurfit Kappa
Arnsley Road, Weldon Industrial
Estate,
Corby, Northants, nw175QW
Tel: +44 153 640 6784

Pouch-Up® — high performance film, convenience and a low carbon footprint.

Three different film structures — Triplex PET MET, Triplex EVOH, and Quadruplex Aluminium.

The entire surface can be printed, ensuring brands stand out on the shelf.

The innovative film and the Vitop® compact tap provide a high oxygen barrier, allowing the drink inside to stay fresher for longer.

Meets the growing needs of on-the-go consumption

3D Lenticulars

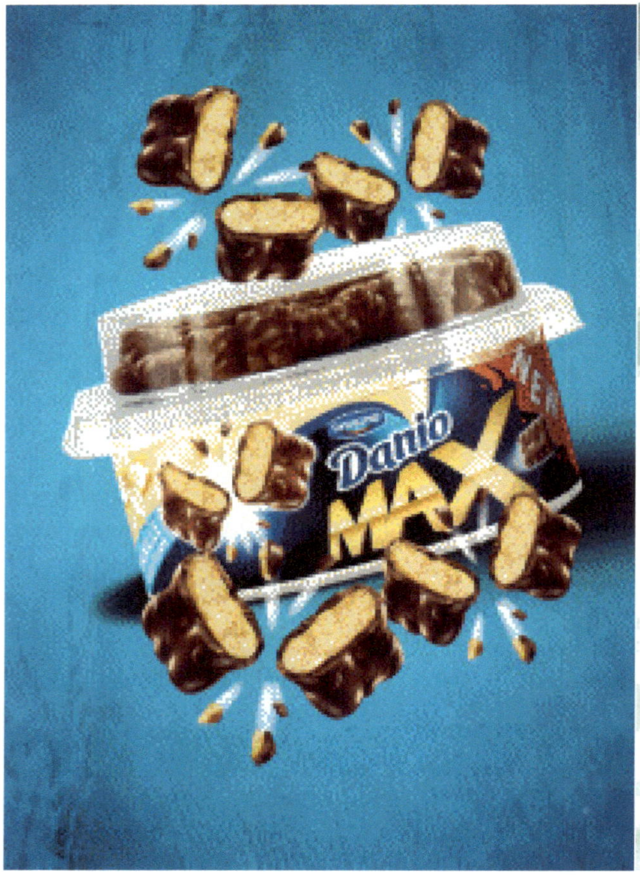

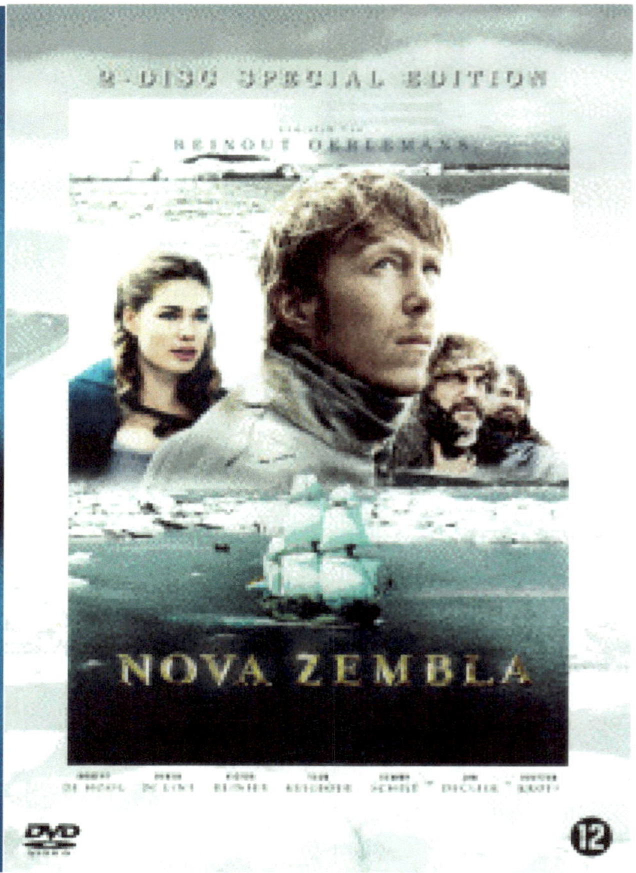

Manufacturer/Designer

Pacur, LLC"3555 Moser St"Oshkosh, WI 54901"USA
Phone: +1 (920) 236 2888
info@pacur.com
Pacur.com

Name : Danone – Danio

Dimensions : width : 670 mm, height : 970 mm

Effect type: Flip, 3D effect – depth

Viewing angle / distance: 49° / 1,2 meter

Material: Lenstar 40 LPI Pacur

Digital printer used: Fuji Acquity

Description: Large format 3D/flip attached to a big display to draw the attention in the shop. Danone had a stand in the shops where the customer could taste the product. The special lenticular display was successfull.

Printing Company: Cosapack nv

Name : Nova Zembla DVD lenticular Box cover

Dimensions : width : 100 mm – height : 150 mm

Effect type: 3D effect – depth

Viewing angle / distance: xx

Material: Lenstar 100 LPI lens and glued on the cover by using robot positioning.

Item description: This cover was made for the Dutch DVD market

Smartas. — bath tissue
we learn as we go

Manufacturer/Designer

Anagrama
Puebla #114 en la Colonia Roma
Norte,
Código Postal 06600
Delegación Cuauhtémoc,
Distrito Federal, México.
T + 52 (81) 8336 6666
hello@anagrama.com

This simple designing is a visual idea that represents brands fun and youthful spirit.

The logo depicts a mortarboard and toilet tissue combination.

The design makes a conscious effort to highlight the paper's bright colours.

Rounded typeface to give the brand a

Choose your cereal

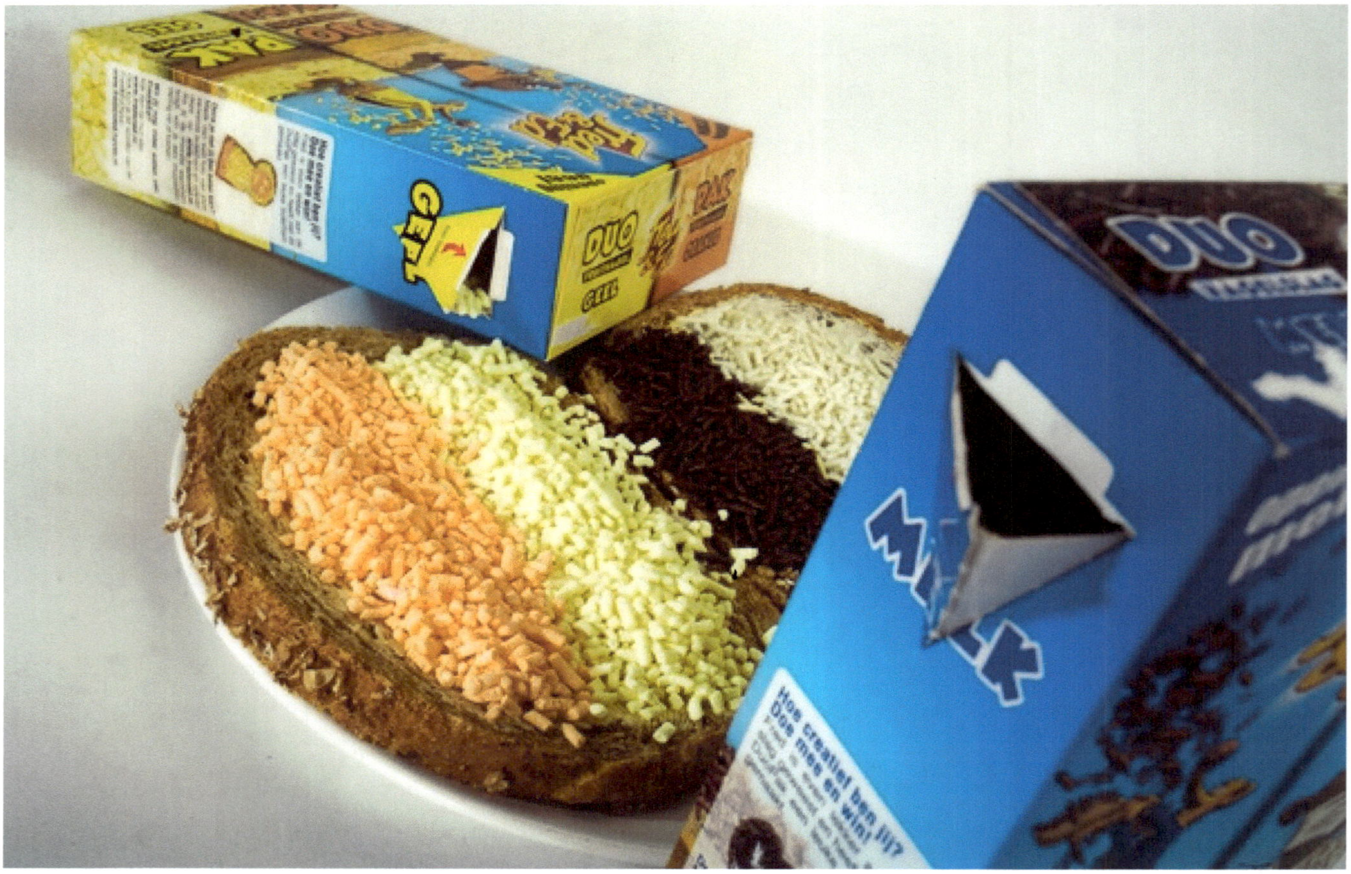

Manufacturer/Designer

MMP Caesar GmbH & Co. KG
31061 Alfeld (Leine),
Fritz-Kunke-Straße 8
Phone: +49-(0)5181 80 08-0
Fax: +49-(0)5181-80 08-88
E-Mail: behrens.alfeld@mm-packag-ing.com

The idea being, to make breakfast more interesting for children. a packaging with two dispenser openings and containing different products.

The two dispensers are an integral part of the design, there is no need for additional dispensers.

The graphics too were very effective showing clearly that the carton contained two different products

The pack was also sift proof to ensure that the contents did not leak and it was clearly shown how the pack worked

"Pull out pouring spouts made the whole product functional and easy to use "

Easy Lift Lid

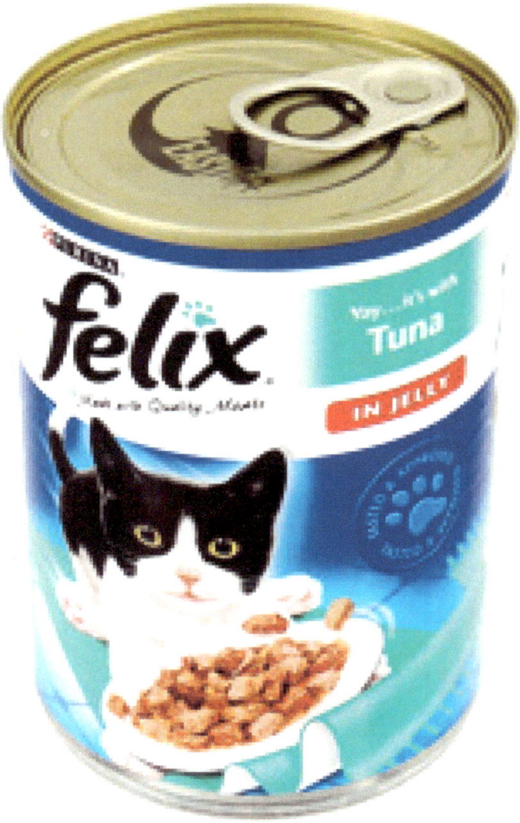

Manufacturer/Designer

Hella Neffati, Marketing Manager;
Tel: +1 215 698 6056;
Email:hella.neffati@crowncork.com
http://www.crowncork.com

Easy-open end now features a more generous gap between the can lid and the ring-pull tab.

This enables greater finger access making the can is easier to open.

From a production point of view these new ends are compatible with existing can line equipment

Easylift easy-open ends offer a simple and important benefit – greater consumer convenience

Improved tab access makes it easier and faster for consumers, including seniors, children and the physically impaired, to open canned food products without using a can opener or other tools.

gogol mogol !

EASY EGG BREAKFAST

■ Outer layer is paperboard

Each egg is individually wrapped in packaging that contains four layers.

■ Second layer is infused with chemicals

■ Smart layer containing water

Membrane

1 Pulling the tab removes the membrane to allow the chemicals and water to come into contact generating heat.

2 The egg is cooked after 2mins, producing a runny yolk, although the heating process will continue for up to 3mins.

3 The user twists off the top half of the pack while the bottom half serves as an egg cup.

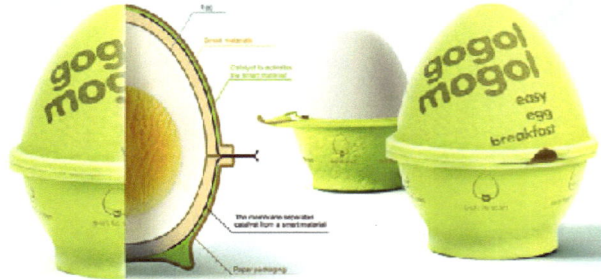

Manufacturer/Designer

Designer: Evgeny Morgalev KIAN
info@kian.ru
Moscow "Ostapovsky, 3, Building 27-29
+7 495 926 09 86

The packaging contains a chemical layer which, when triggered, generates heat and cooks the raw egg in just two minutes.

The outer layer is made from the sort of paperboard traditionally used to make egg boxes.

Beneath this there are three more layers. One is infused with calcium hydroxide and other chemicals, and the other is a 'smart layer' containing water.

'It uses calcium hydroxide and water, so when the components come together a large amount of heat appears.

'Under the cardboard layer is a catalyst and a membrane, which separates the catalyst from a smart material.

"When you pull out the membrane by stretching a tag, the chemical reaction between the catalyst and a smart material

Instant soups in collapsible cups

Manufacturer/Designer

Nestlé Product Technology Centre in Singen (Hohentwiel), Deutschland

Collapsible-cup packaging for instant soups.

A combination of a flexible stand-up pouch that contains the product and a folding paperboard part that turns into a convenient cup upon squeezing the top rim of the pouch.

Speciality of the packaging for "Moment Mahl" soups is that the soup bowl is already integrated in the pouch.

The consumer just has to tear off the top edge of the foil and squeeze the surrounding paperboard ring at the marked locations until the cup clicks, creating a stable soup terrine.

Then pours boiling water into the cup, stirs, waits three minutes, after which the soup can be enjoyed.

Hot Sandwich in Flight

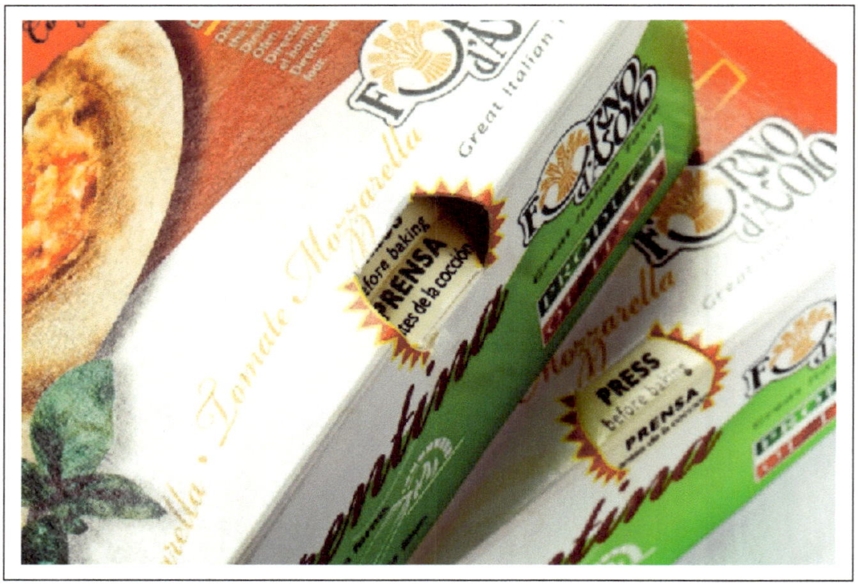

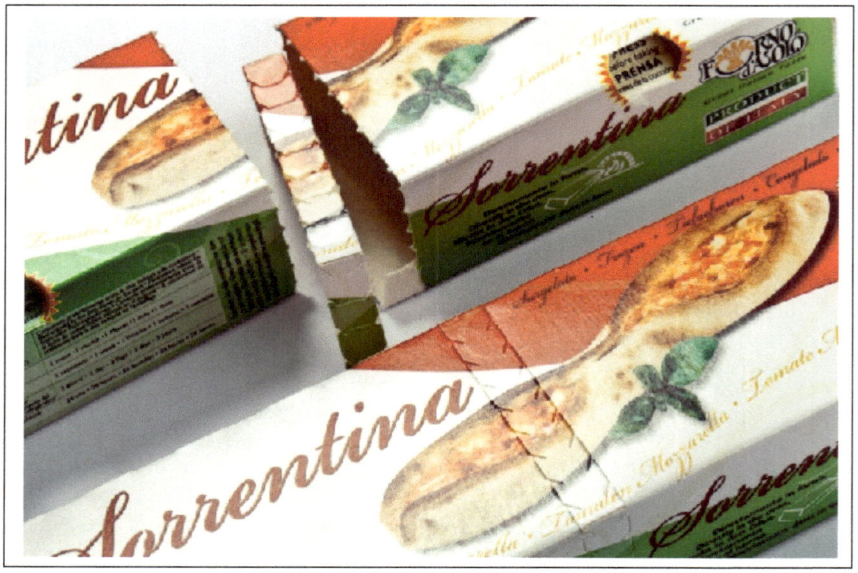

Manufacturer/Designer

BOX MARCHE SPA
Via San Vincenzo,
67-60013 Corinth (An) Italy
Tel +39 071 79789.1
Fax +39 071 7978950
info@boxmarche.it

Special folded carton for catering on aircraft: a product which could be heated in an oven together with its deep-frozen content without burning.

Prior to heating, the two specially designed areas on the sides are pressed inwards using the thumbs to avoid overpressure in the oven.

After heating, only the centre strip needs to be torn open which divides the box into two halves.

This allows eating a hot sandwich without burning or soiling. The carton is also subjected to special treatment (Jazz heat treatment) which creates the barrier for heating in the oven."

Two small push perforations had been added to allow air to escape safely whilst being heated but without the risk of contamination during shipping

Rattle Tattle

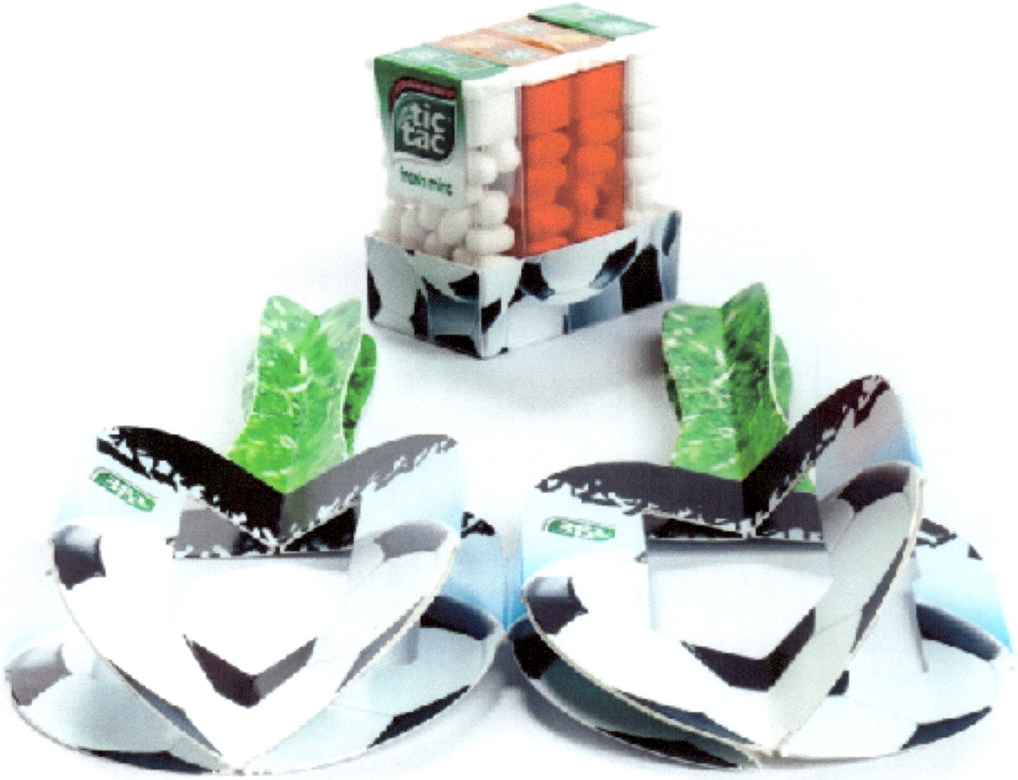

Manufacturer/Designer

STI Line Ltd. Pentland House Saracen
Close Gillingham Business Park
GB-ME8 0QN Gillingham
Tel: +44 1634 377590
Fax: +44 1634 377560
service.uk@sti-group.com

The sound of packaging"Packaging for the football world championships 2014 in Brazil that will appeal to fans."

The rhythm instrument Maraca, which is typical for Brazil. Inspired a fan rattle for tic tac,

The shaking of a pack of tic tac and the typical sound of a rattle could be combined perfectly with the idea of a Maraca. Each rattle included two packs each of tic tac Mint and Orange. By using cartonboard,

The individual tic tacs could be removed from the packs without tearing the carton apart.

Additional added value was gained as the sound could be altered by changing the filling depth. In addition the design was well conceived as the pack was

On the go fresh Coffee

Pull string to open
for pour spout

Pour spout and
coffee exit

Filter compartment
with 25 g of freshly
ground specialty coffee

PE-coated paper

Size is like A5 paper
and thickness is 1 cm

Weight 45 g

Manufacturer/Designer

Coffeebrewer Nordic A/S
Kasmosevej 3,
DK-5500, Middelfart,Denmark
Tel: +45 63 400 124
Fax: +45 63 400 125
contact@growerscup.com

Coffee brewer for on-the-go"disposable coffee brewer, that works right inside its own stand-up pouch.

Inside the pouch is a filter with 26 g of freshly ground coffee. To brew 3 cups of coffee, all you need is to open the pouch, pour ½ litre of hot water into it, and let it brew for 5-8 min."

The filter will effectively separate the brewed coffee from the grounds and

when you have served the first 1½ cup the remaining coffee will be under the filter and the brewing process will stop.

This means that the coffee will not go bitter over time like in a French Press.

Reclosable and Tamper Evident Dispenser Carton

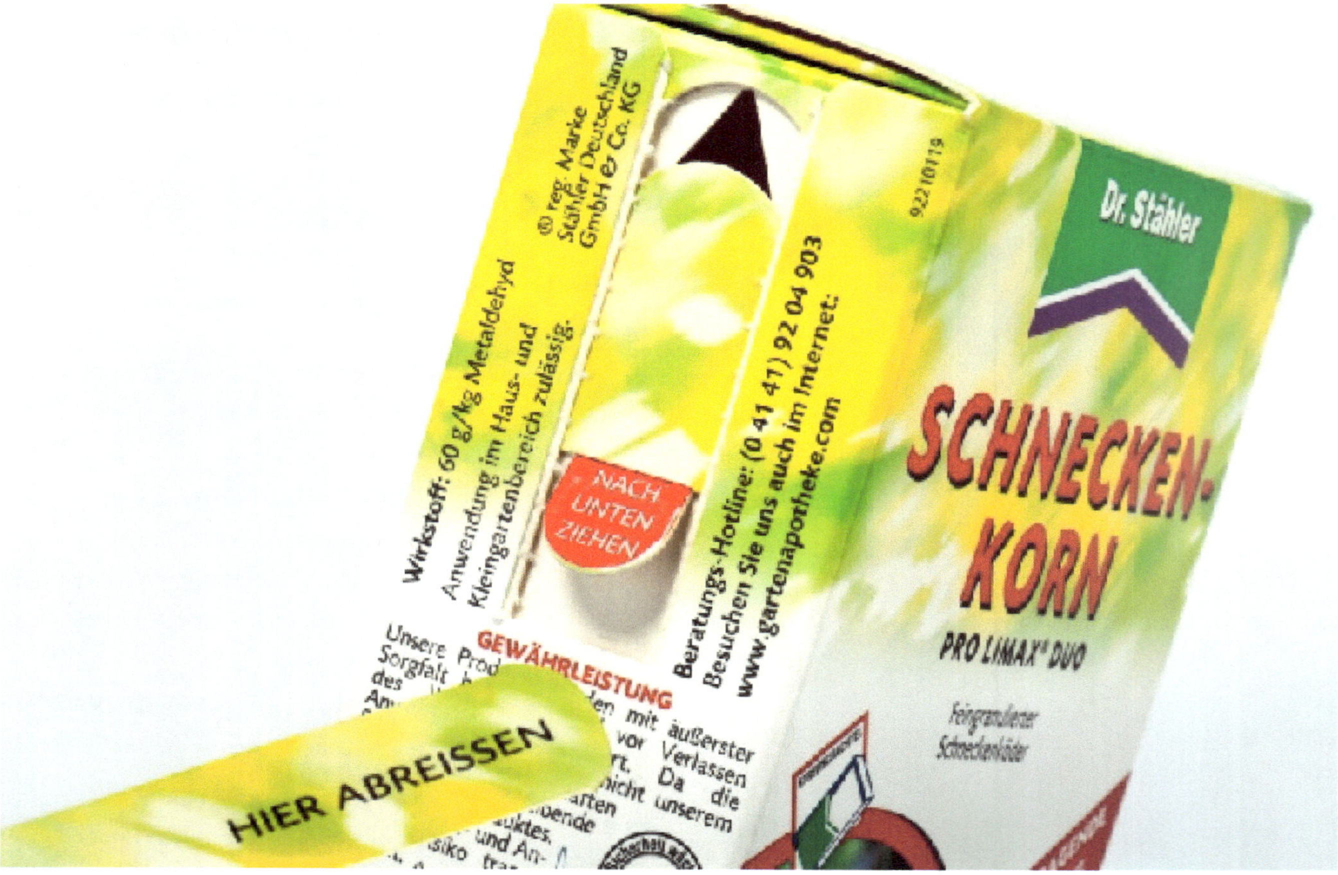

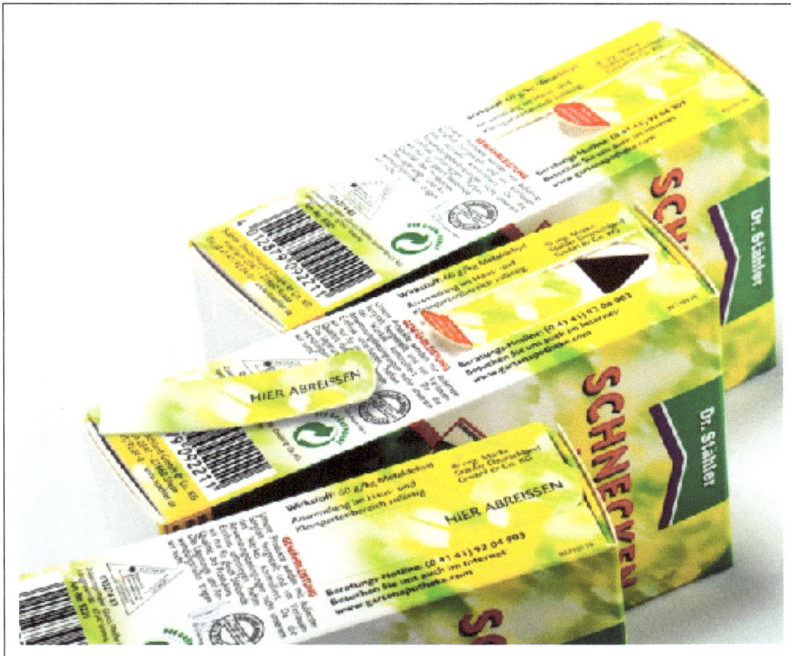

Manufacturer/Designer

Karl Knauer KG
Zeller Straße 14.
D-77781 Biberach/Baden
Tel: +49 (0) 7835 7820.
Fax: +49 (0) 7835 3598.
Email: info@karlknauer.de

The cartonboard folding box with re-sealable dispenser opening and tamper-evident closure.

It was developed further and optimised for the slug pellet product.

This solution guarantees the consumer both the originality of the product as well as easy handling.

After use, the packaging can be resealed firmly which fully protects the slug pellets. ""Made of a mono-material and can be erected, filled and sealed in a fully automated process.

Design contained a tamper evident seal which once opened had behind it a sliding mechanism for dispensing

Extremely simplified recycling process which makes this product highly sustainable

BREAKTHROUGH

Dosage Stick Packs with CR Features

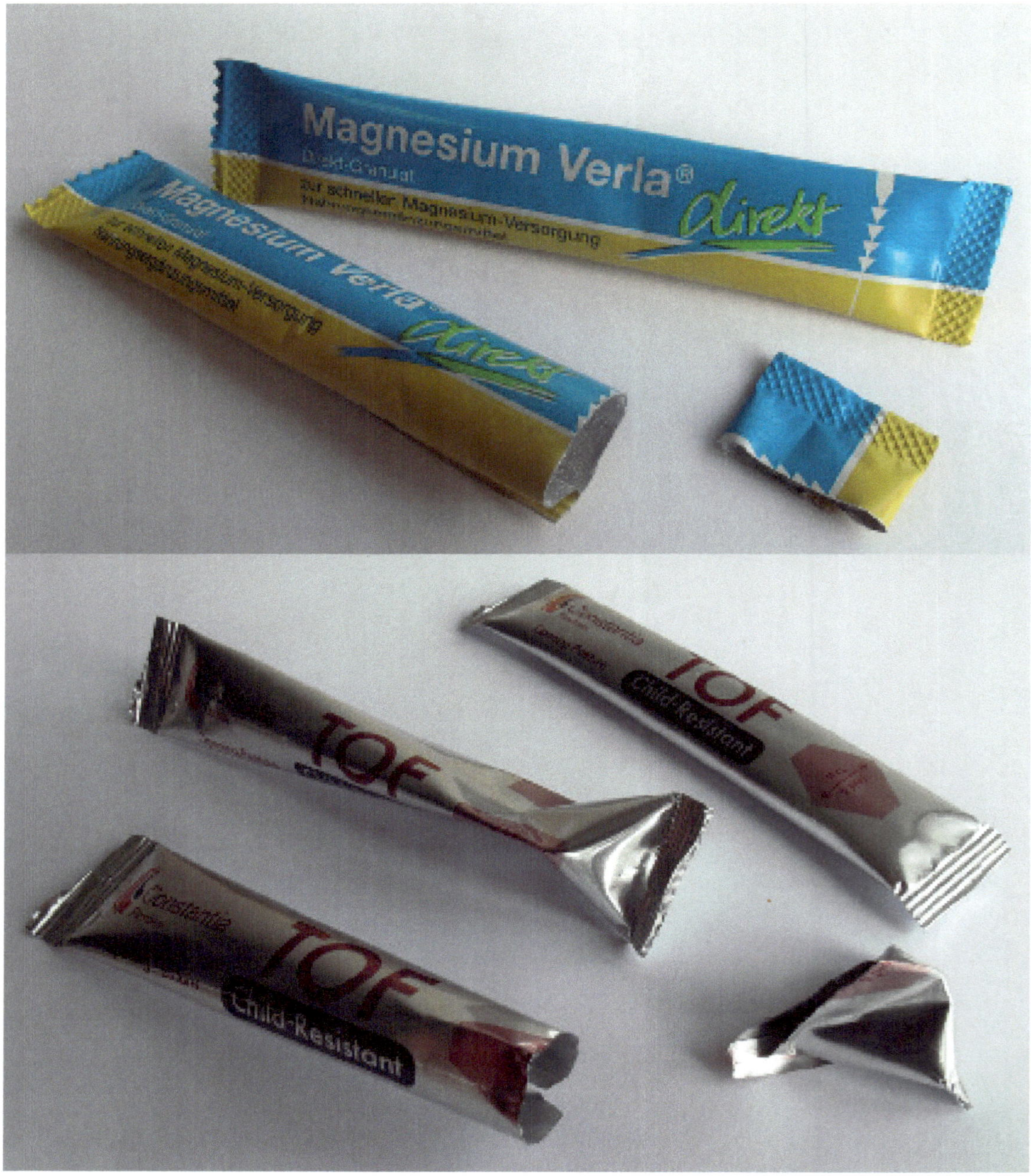

Manufacturer/Designer

Constantia Flexibles
Rivergate, Handelskai 92
1200 Vienna
Austria
T +43 1 888 56 40 1000
F +43 1 888 56 40 1900
office(at)cflex.com
www.cflex.com

Stick packs with easy and clean opening for dosage stick packs

The Laser Perforation Opening Aid can be opened quickly and simply and is clearly marked by arrows at one end of the stick.

Using the PET/Alu/PE-LD easy-tear laminate and laser perforation allows 100 per cent of the aperture to be opened.

The stick pack with Opening Aid TOF introduces a micro-perforation in a defined area of the PET layer of the laminate.

The device is also suited to for pharmaceuticals, personal care or food.

Both forms of opening come in child

On the Go salad!

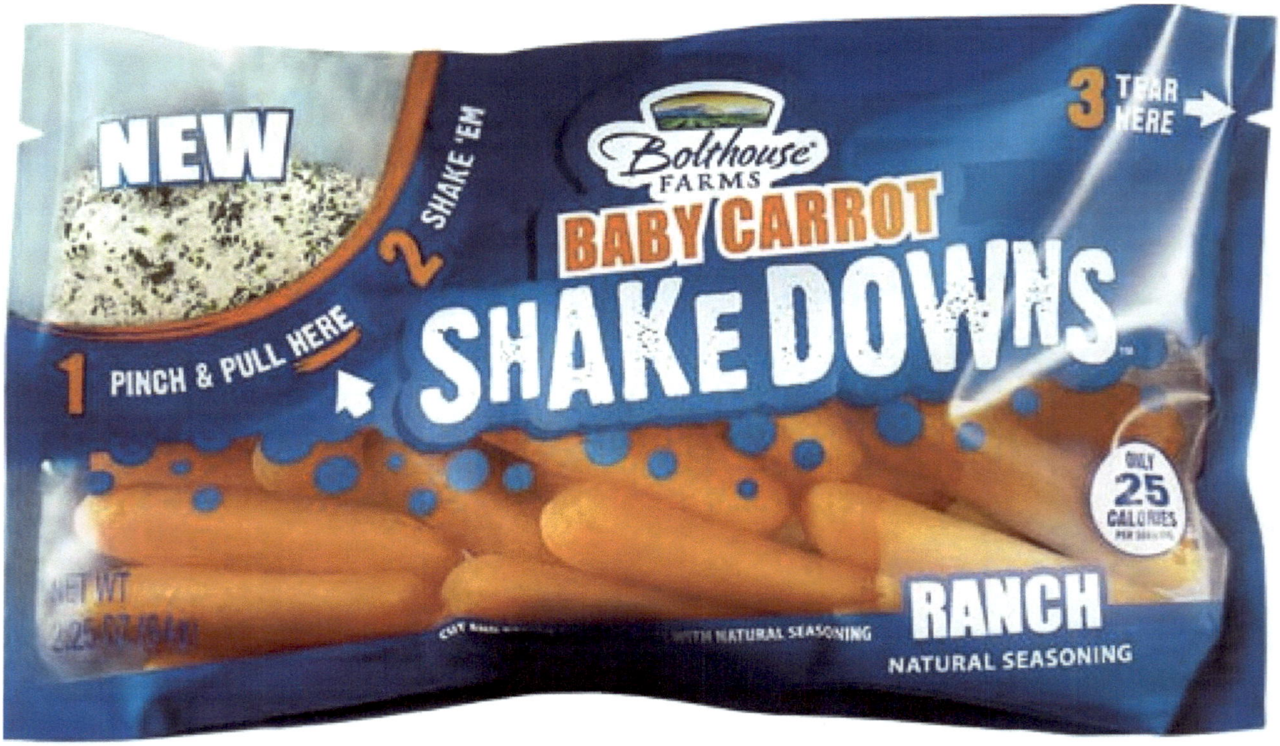

Manufacturer/Designer

EMARALD PACKAGING
Todd Somers "phone:
510.429.5700 "fax: 510.429.5715
sales@empack.com

Package has changed the landscape in the unit level segment of the fresh cut produce market.

The interactive nature of the package brings consumers into the role of participant in preparing their vegetables in a simple and fun way.

This is done in a single package combining wet and dry products without compromising shelf life of the carrots.

The consumer breaks open the packet to disperse the seasoning over the carrots, then shakes to coat them.

Confectionery pack with magnetic closure

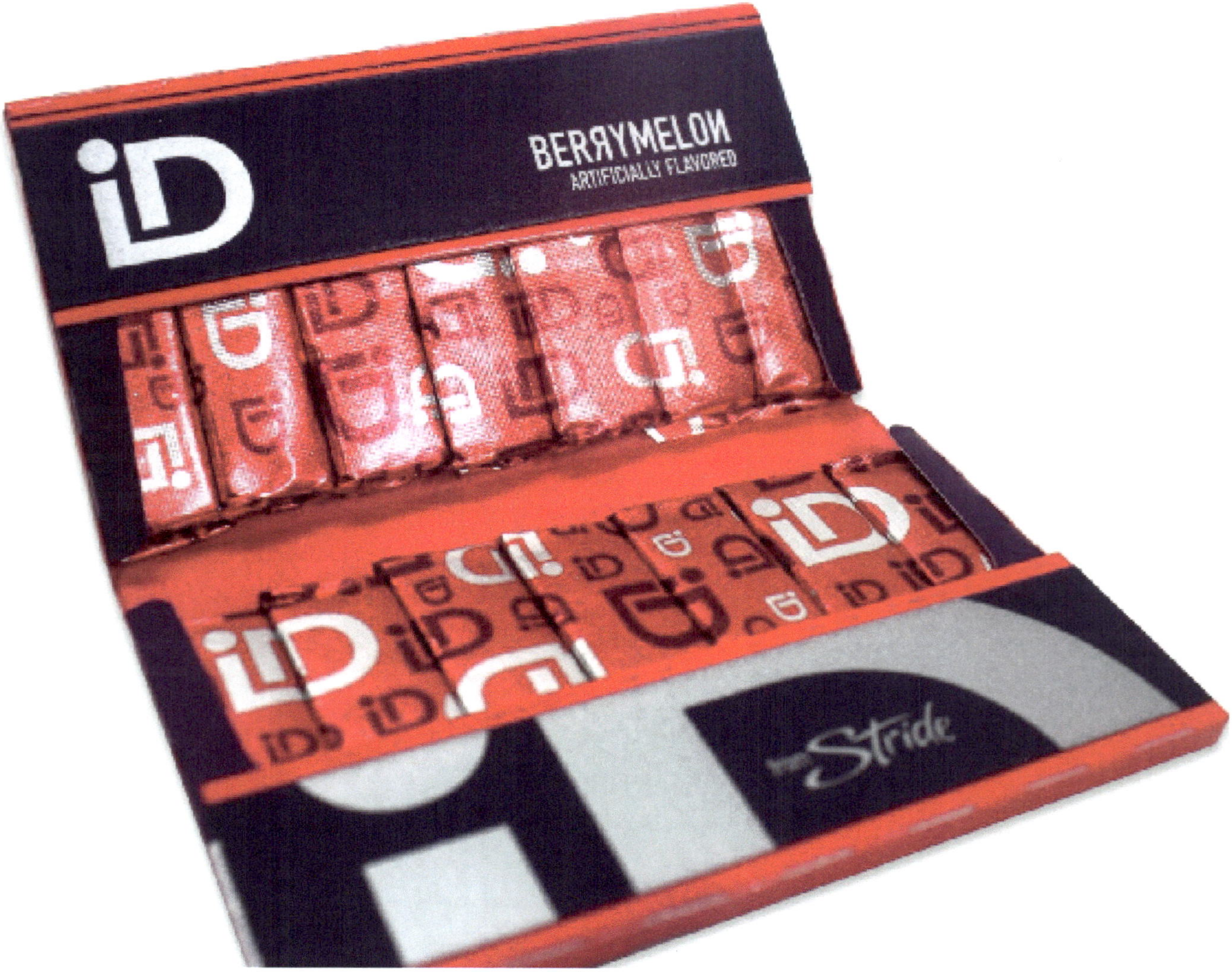

Manufacturer/Designer

ASG (AGI Shorewood Group)
400 Atlantic Street
14th Floor
Stamford, CT 06921
northamerica@asg-worldwide.com

"The folding packs featuring a new type of magnetic closure.

Considered "first of its kind" and could be a "game changer" in the highly competitive confectionery industry."

The innovative magnetic closure helps easy open and close of the carton without wear & Tear

This new process features a coating which is applied during printing on the outer edges of each side that easily come together and pull apart.

Gum, is wrapped in foiled paper, Clear film overwrap provides tamper resistance.

This specific package with magnetic coating was approved by the FDA, and that "there is no direct food contact

Other Application: Cosmetics & Personal

UV protective shrink sleeve

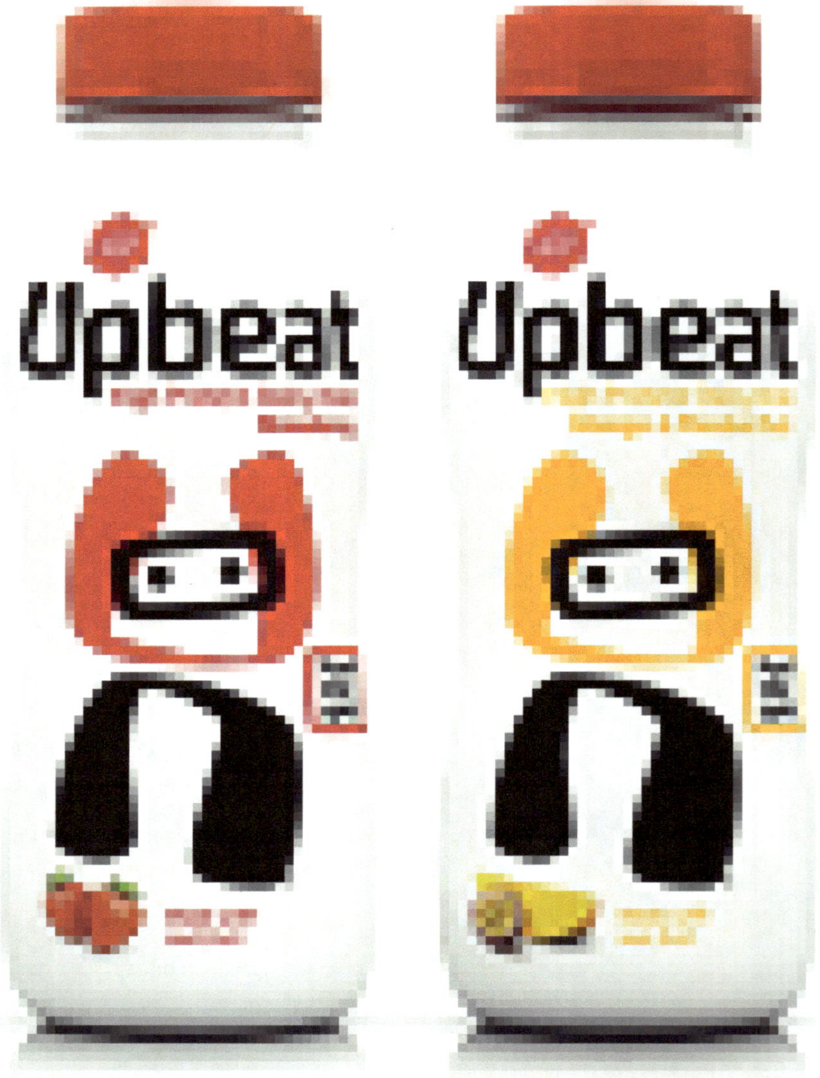

Manufacturer/Designer

Hohenems
CCL Label GmbH
Riedstrasse 2, 6845 Hohenems,Austria
Phone: +43 5576 7111 0"Fax: +43 5576
7111 18
Martina Harrer, mharrer@ccllabel.at
www.ccllabel.at

Special decorative incorporating UV protection sleeves to protect a high protein dairy against UV light degradation.

Packaging projects a premium image and distinguished it from traditional UHT drinks.

A custom-designed PET bottle with a tapered waist. The sleeves, which wrap tightly around the contours of the bottle

The sleeve is printed with a UV absorbing lacquer.

Are flexo printed in nine colours with a matt silk finish that is smooth to the touch and further reinforces the brand's

TE cap for Icecreams

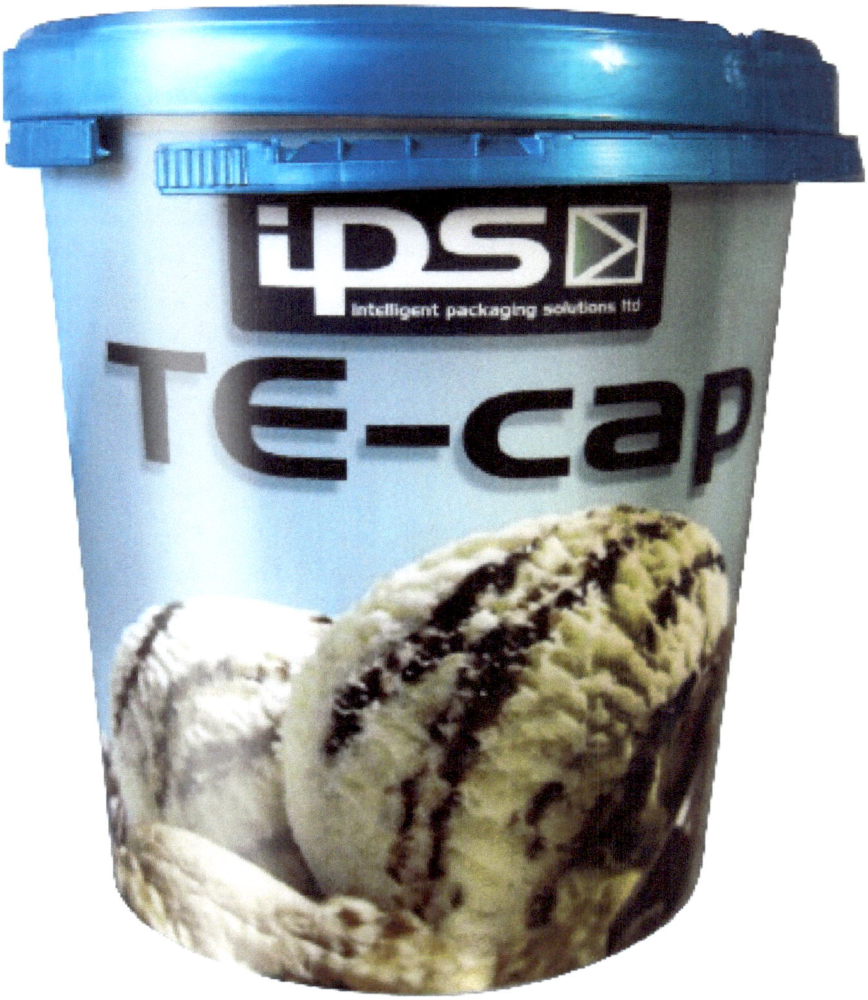

Manufacturer/Designer

Intelligent Packaging Solutions
Stephenson Road
Stephenson Industrial Estate
Washington
Tyne and Wear
NE37 3HR
Tel: 0845 0020039
Fax: 0191 415 6501

A new innovation, the TE-cap is a combination of a lightweight paper pot and a deep-freeze grade polypropylene lid.

The patented lid design provides a tamper evident seal delivering product security, with easy opening and re-closure for consumer convenience.

This innovative lid means that now even paper containers can be made Tamper Evident.

The packaging features an injection moulded break-off tab and skirt which eliminates the need for foil or membrane sealing and provides instant evidence of any interference.

This is a convenient way to assure the consumer that package contents

Peelable Soup pouches

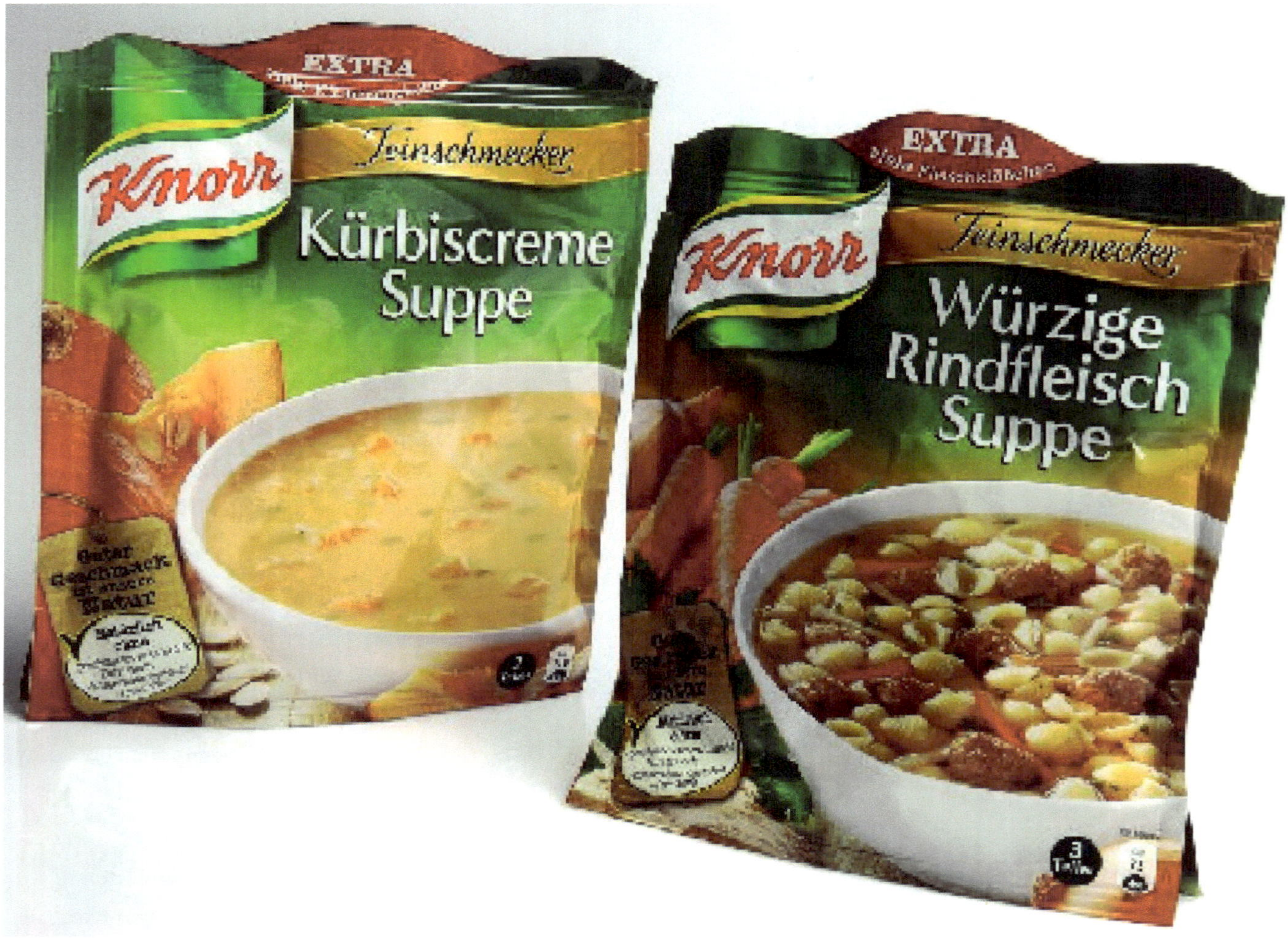

Manufacturer/Designer

Constantia Flexibles
Rivergate, Handelskai 92
1200 Vienna
Austria
T +43 1 888 56 40 1000
F +43 1 888 56 40 1900
office(at)cflex.com
www.cflex.com

A new format in dry food applications

Due to the combination of two flaps and easy peel opening the consumer can open the pouch in a very convenient way

No additional equipment required

Additionally the opening offers the end user a clear look into the pouch and its contents.

The pack does not get split into two pieces when opening —the branding remains intact and full readability

The pack can also be disposed of more conveniently in one piece.

The three layer laminate is reverse-printed PET/alu/ peelable LDPE

Retains the full barrier and machinability functionality of the four layer material used by conventional pouches.

IML with barrier layer

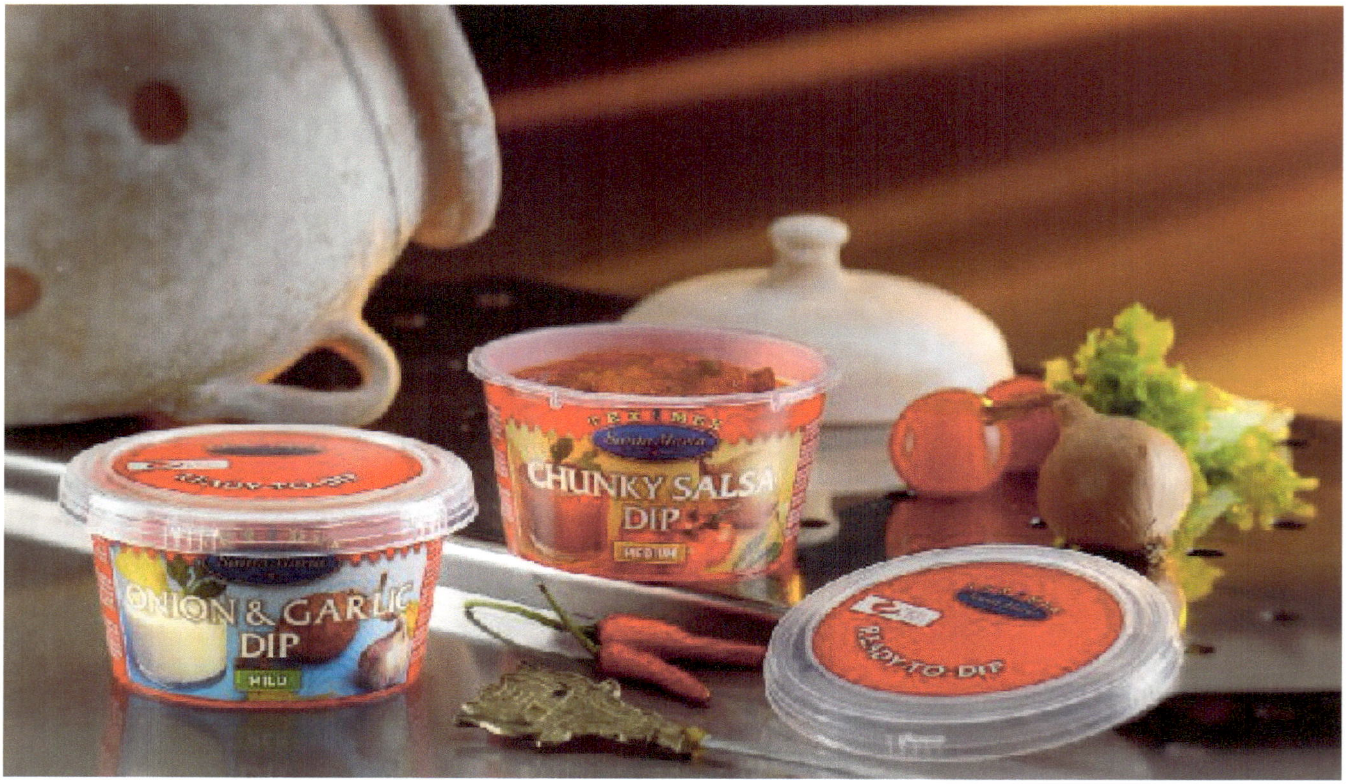

Manufacturer/Designer

RPC Superfos
RPC Superfos a/s
Spotorno Allé 8
2630 Taastrup
Denmark
Phone +45 5911 1110
Fax +45 5911 1180
Email superfos@superfos.com

In-mold label has barrier layer injection molded of polypropylene offers In-Mould-Labelling with barrier properties.

A layer of ethylene vinyl alchohol (EVOH) in the PP label provides gas barrier properties that prolong shelf life.

The stacking facility of the SuperLock package makes it easy to handle in connection with the filling process.

With in-mold labelling, subsequent labelling is eliminated and this makes the filling process more efficient.

Convenience products can stay on shelf twice as long - or even three times longer - in SuperLock than in reference

New layout for fluted cartons

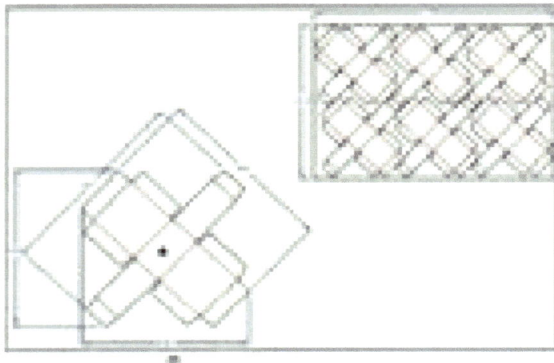

Manufacturer/Designer

Smurfit Kappa
Ballymount Road,
Walkinstown, Dublin 12
Tel: +353 1 409 0000

The product designer at Smurfit Kappa went against convention and rotated the direction of the corrugations on the E-corrugated board.

The twisted arrangement of the cuts on the punching sheet resulted in a considerable improvement in the layout.

This led to customer savings of 20% in the packaging. In addition, the improved technical characteristics meant that the diagonal corrugation achieved a roughly 13% higher BCT value (stacking compressing resistance, measured by a Box Compression Test).

This improves the machine opertaions - with improved flatness and uniform bending resistance of the grooves, the blanks are cut more smoothly and thus processed faster.

Sustainable Cupholder for Yoghurt and Museli

Manufacturer/Designer

WS Quack & Fischer GmbH
Heinz Eicker
Management
Phone: +49 (0) 21 62/938-0
Fax: +49 (0) 21 62/938-274
heinz.eicker @ wsqf-verpackungen.de

A sustainable idea of placing the cups on top of each other.

This saves on material, requires less space in the refrigerated section and handling in the car is also very easy.

The colours symbolise the freshness of the product and the "Cupholder Müsli & Yoghurt" needs extremely little space on the refrigerated shelf.

The new shape and simplified handling will attract consumers' attention

A comparatively simple cartonboard construction holds two pots inside safely and securely

TECHNOLOGY

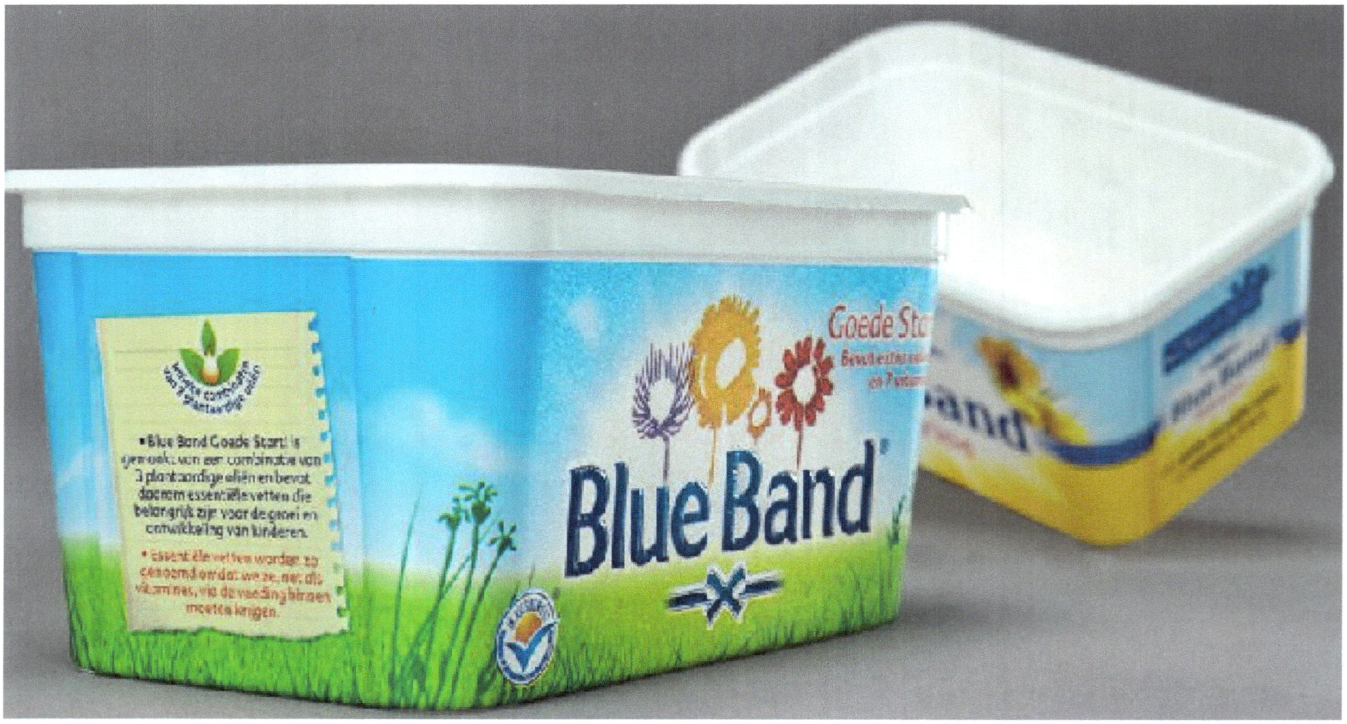
Blue Band

Manufacturer/Designer

Trexel, USA; PACCOR International GmbH, Germany; Britton Decorative, France

3D IML combines MuCell technology with IML label technology to offer a unique visual appearance and soft touch effect to packages,

Providing differentiation without changing the container shape. Multi-Sensory Experience

The 3D in-mold label is processed with an injection-molded tub made of expanded polypropylene to incorporate selective masking, creating areas without adhesion.

The MuCell process controls the introduction of nitrogen into the polymer.

Where there is adhesion between the label and the polymer, the nitrogen migrates through both structures.

Where there is no adhesion (by design), the nitrogen accumulates between the polymer substrate and the label. The process creates bubble patterns or Braille

Bombay Sapphire

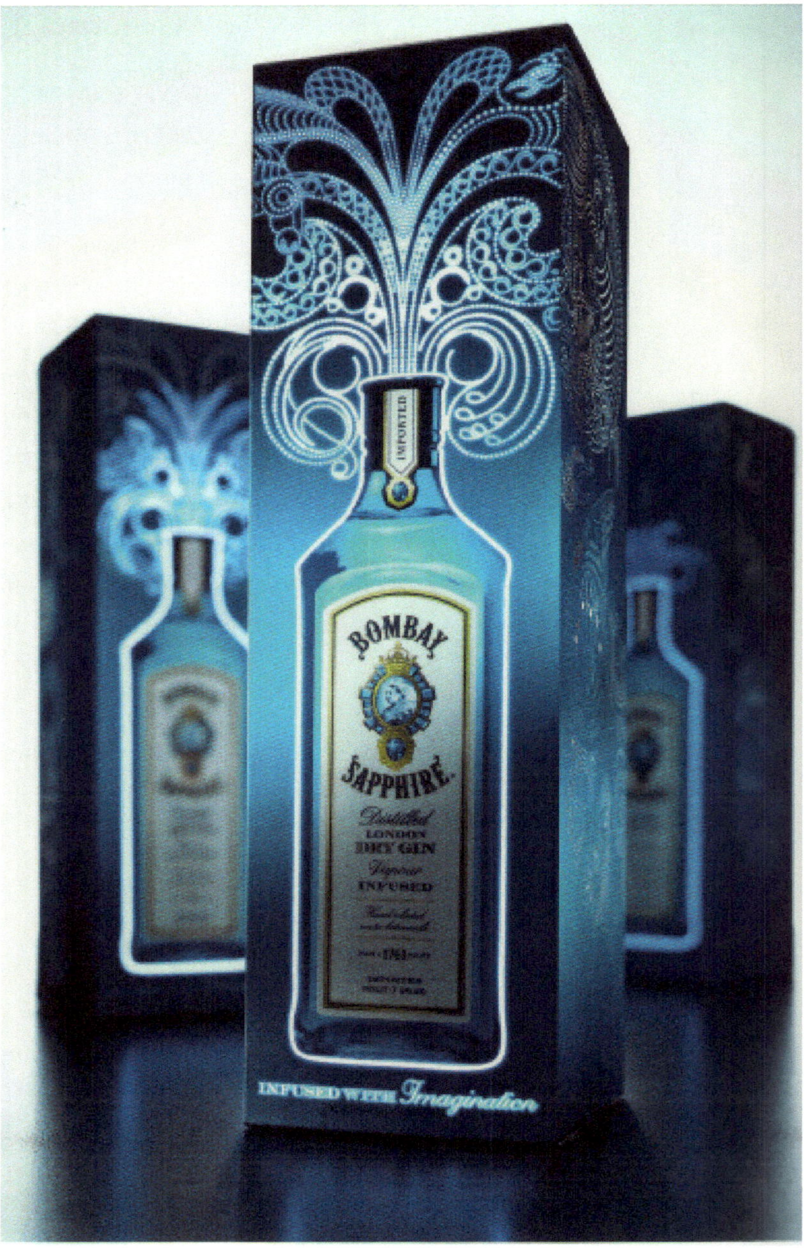

Manufacturer/Designer

Karl Knauer KG
Zeller Straße 14.
D-77781 Biberach/Baden
Tel: +49 (0) 7835 7820.
Fax: +49 (0) 7835 3598.
Email: info@karlknauer.de

This folding carton is probably the world's first packaging with printed, actively illuminated surafces.

Implemented with the protected technology "HiLight — printed electronics".

The illuminated animation on the front of the packaging is triggered by an innovative and intuitive mechanism as soon as the packaging is held in hand.

The illuminated scene is set in five consecutive stages from bottom to top.

This packaging defines new standards for the presentation of a brand at the POS with a degree of attention which has

FMCG INNOVATIONS

FMCG INNOVATIONS

Manufacturer/Designer

bluemarlin Asia
80A & 82A
Tanjong Pagar Road
Singapore 088501
T: +65 6222 6503
Email: singaporecrew@bluemarlinbd.
com

Soft drink in aerosol format, targeted at teens.

Turbo Tango features "nitro fuelled" aerosol technology, to deliver a foamy blast of orange and a totally new drinking experience.

Edgy and disruptive bottle to engage with its target audience

The world's first soft drink dispensed through an aerosol container and has licensed rights to the patented

Nor Cell (written in next sheet)

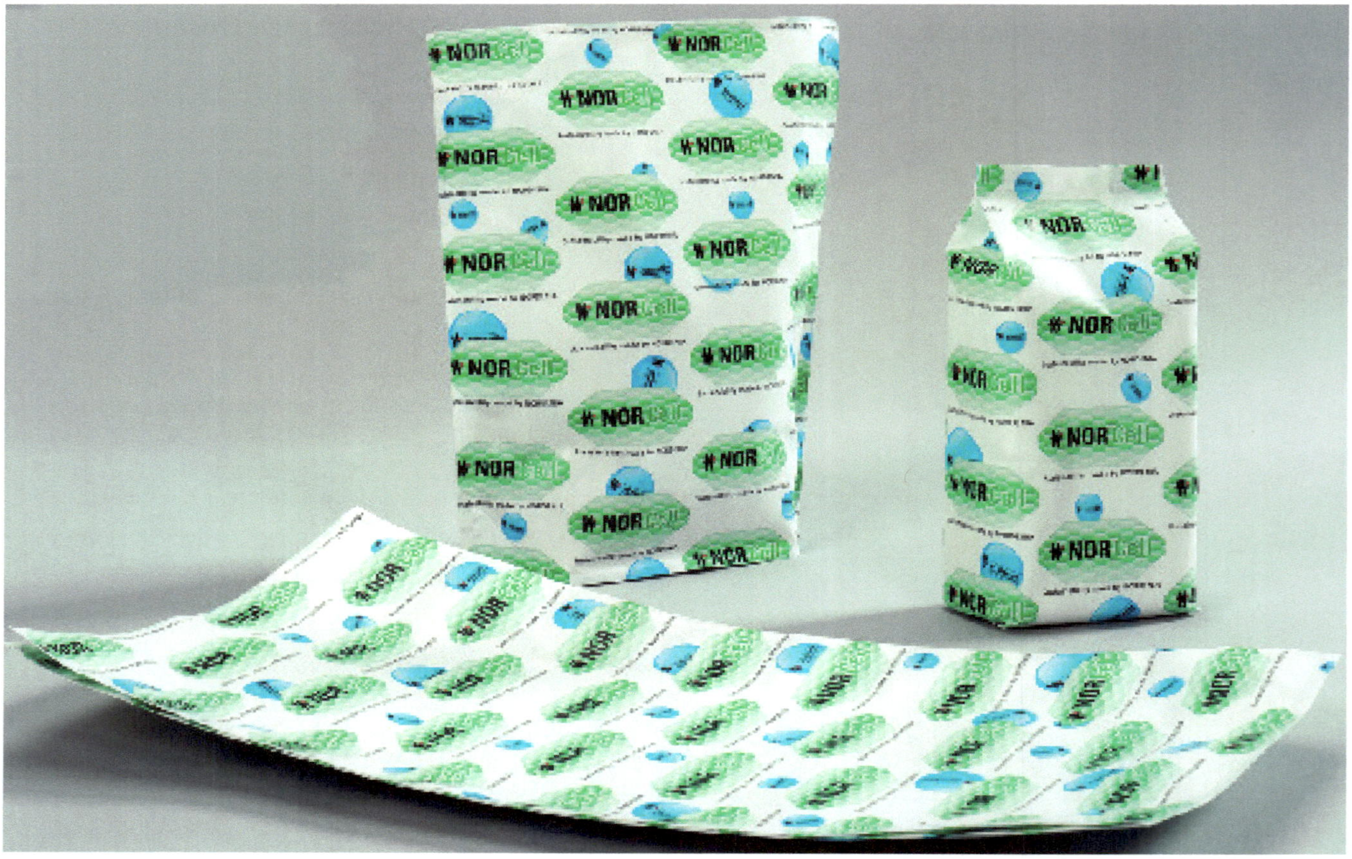

Manufacturer/Designer

NORDENIA INTERNATIONAL AG
Airport Center am FMO
D-48268 Greven, Germany
Phone: +49 (0)25 71 / 91 91 40
Fax: +49 (0)25 71 / 91 91 91
cord.witkowski@nordenia.com
www.nordenia.com

Nor®Cell — the innovative, sustainable and resourceful way of packaging. the Nor®Cell technology in flexible packaging.

The highlight of the Nor®Cell technology: the weight of flexible packaging is greatly reduced using a physical, controlled foaming process without reducing the film thickness.

This "lightweight construction film" weighs up to 40 percent less than conventional film. The material is therefore intelligently and efficiently used.

Surface structures with unique haptics of the film can be achieved through the patented foaming technology.

At the same time, packaging made with Nor®Cell is characterized by a high puncture resistance, excellent sealing ability and great printability.

This therefore meets all of the customer requirements for high quality, flexible product packaging.

Hgh quality packaging is currently being produced

FMCG INNOVATIONS

QR codes

i'm lovin' it
Simple
ENJOYMENT
MORE INFO
FREE
WI-FI

HEINZ
OUR TURN
TO SERVE
SUPPORT OUR
VETERANS
TOMATO KETCHUP

OUR TURN TO SERVE

Manufacturer/Designer

McDonald's

New packaging is designed to engage with customers in relevant ways.

Customers always want to know more about the food they are eating and that is easy by putting this information right at their fingertips.

The bags and cups feature QR Codes providing access to nutritional information.

Provides consumers with information to help them make informed choices.

Text has been translated into 18 different languages.

A blend of text, illustrations and a QR code will deliver interesting facts about the brand and make nutrition information easily accessible from

VALUE ENGINEERING

Manufacturer/Designer

Constantia Flexibles
Rivergate, Handelskai 92
1200 Vienna
Austria
T +43 1 888 56 40 1000
F +43 1 888 56 40 1900
office(at)cflex.com
www.cflex.com

A new lidding concept, which reduces the lid material Die Cut Lid 2020.

A co-extrusion coated lid is composed of a thin aluminium layer of 20 micron and a proprietary co-extrusion coating.

Die Cut Lid 2020 is sealable against PP-cups and is available embossed or unembossed.

Constantia claim it is the first die cut lidding product to use only 20 micron foil.

clearly shows how advances in material technology are making possible better sustainability without loss of

Seallite Lid (As written in the next sheet)

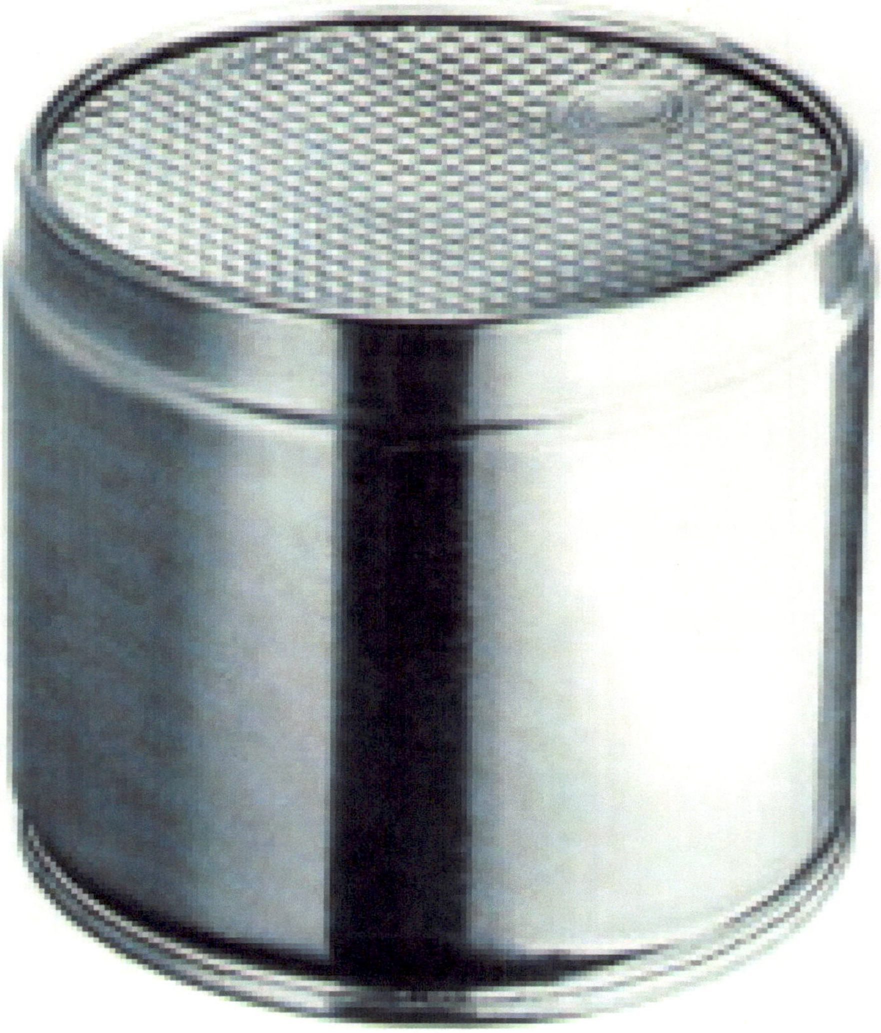

Manufacturer/Designer

Ardagh Group
Merthyr Tydfil
CF48 1PQ Wales
T: +44 (0) 1685 354 400
F: +44 (0) 1685 721 468
http://www.ardaghgroup.com

The SealLite® lid with its preservation and easy open properties.

Used for dried food products the novel lid is produced by cutting out a pre-shaped membrane from a coil of laminated alufoil.

While at the same time the upper edge and the pin of the shaped can is being heated.

The foil lid is then sealed directly onto the necked cans and sent to the customer where they are filled and closed upside down.

When filled and stacked, pressure is applied only to the outside rounded seam; no force is exerted on the SealLite® membrane.

25%
More Soup
See Side Panel*
RIGHT FOODS
DR. McDOUGALL'S
Vegetable
Gluten Free
All Natural Soup
No Bisphenol A
NET WT. 18.0 OZ. (510g)
Vegetable
All Natural Soup

Manufacturer/Designer

Tetra Pak Ltd.
Bedwell Road,
Cross Lanes LL13 0UT,
Wrexham
Phone: +44 1978 834 000
Fax: +44 1978 834 001

Tetra Recart aseptic retort package for its full line of ready-to-serve chunky soup varieties.

The Tetra Recart package saves significant space, weight, and logistical costs throughout its entire life cycle.

The package is said to require 36% less packaging by weight than that of steel cans, and reduces transportation costs.

One truck of empty Tetra Recart cartons is equal to nine trucks of empty steel cans.

The Tetra Recart package is also made mainly out of paper, a renewable resource.

When filled in the Tetra Recart, chunky

R - Laminates

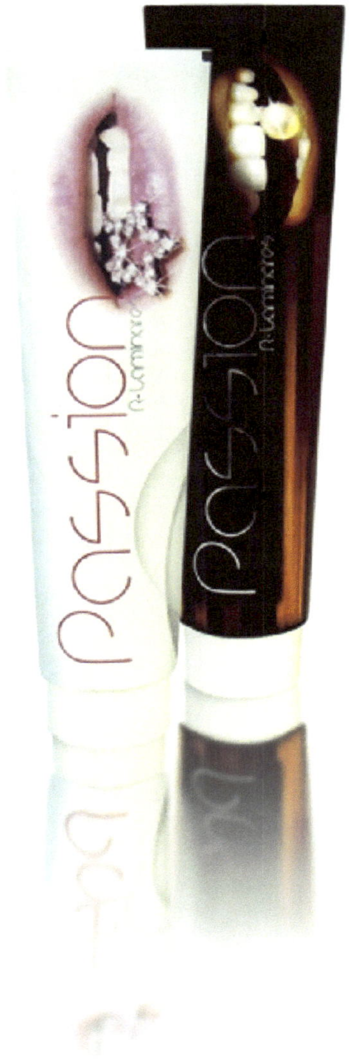

Manufacturer/Designer

Huhtamaki Ronsberg
Huhtamaki Deutschland GmbH & Co KG
Heinrich Nicolaus Strasse 6
87671 RONSBERG ALLGAU
Germany
Tel: +49 - (0) 830 6 770
Fax: +49 - (0) 830 6 772 26
Email: flexibles(at)de.huhtamaki.com

A new laminate for the production of tubes conserve valuable resources.

Compared with a standard laminate R-laminate reduces the thickness and weight of the material between 25 and 45 per cent.

Currently laminates have 250 and 300 micron thickness but R-laminate is between 160 and 185 micron for a toothpaste application

From 500 (extruded) or 400 (laminate) to 225 micron for tubes containing cosmetic products.

In production terms the thinner laminate means saving both material and transport costs.

Give the customer both ecological and economic advantages without loss of quality

FMCG INNOVATIONS

Manufacturer/Designer

Boulder, CO
a division of Boulder Brands

Companies has been switching over from 4.5-in.-dia round polypropylene tubs to 4-in.-wide square PP tubs.

square tubs could free up valuable refrigerated warehouse storage space as well as optimize in-store shelf display space.

The switch can increase warehousing efficiencies by up to 60% and shelf efficiencies by 50%.

Stacks 60 square units on a refrigerated shelf in the same space where only 40 round units could have fit and also increases brand space

A study conducted for Boulder Brands by sustainability consulting firm Renewable Choice Energy found that the new square tub design can reduce greenhouse gas emissions associated with the products' packaging and distribution and the retailers' energy usage by 18%.

The Smart Balance 15-ounce packaging could reduce annual carbon dioxide

Manufacturer/Designer

CP Flexible Packaging
15 Grumbacher Road
York, PA 17406
Phone: 717-764-1193
Fax: 717-764-2039

The new packaging for these products reduced the use of materials by 65%.

The more compact package gives 25% more product space. Resulting in reduced transportation costs and a smaller carbon footprint.

Package features a horizontal flow wrap package made of a film lamination with a re-sealable pressure-sensitive seal instead of a zipper, easy to use, tamper evident increased barrier for longer shelf life and, with a reverse printing process, outstanding shelf appeal.

Traditional closures are replaced with a pressure-sensitive Sealstrip® closure that provides excellent resealability, ensuring that the product stays fresh.

Removal of the tray component enables 65% less material to be used.

Light Weight Bottle

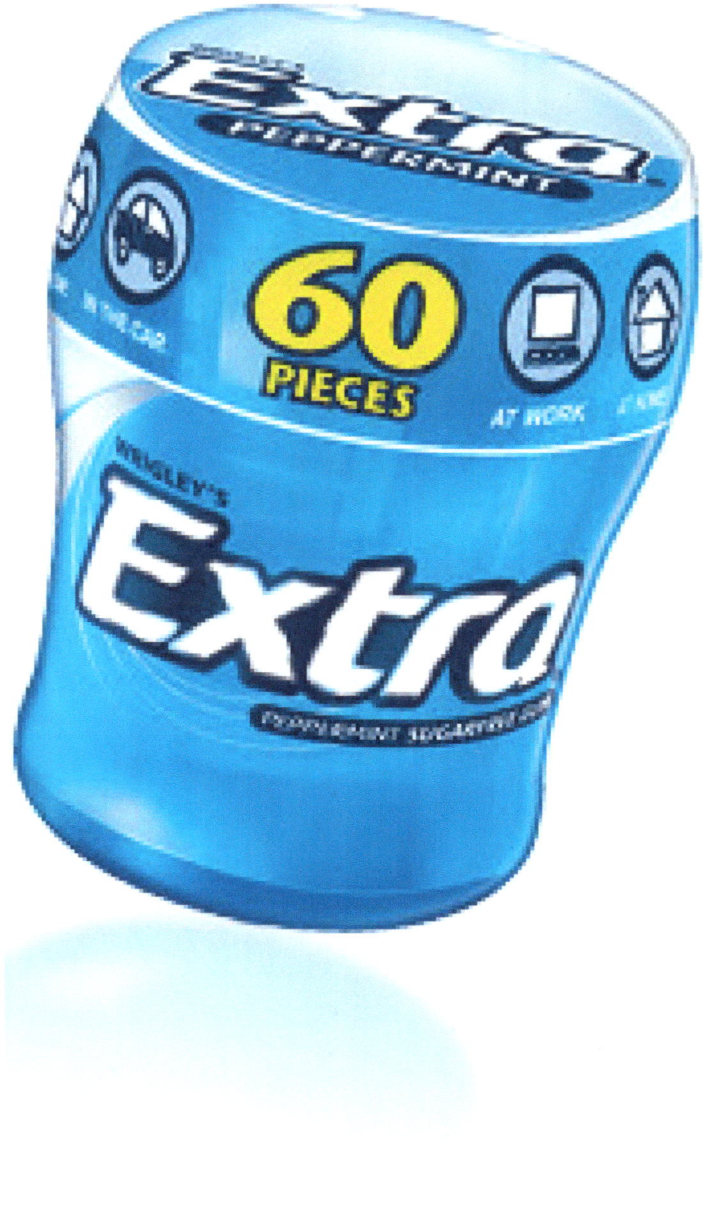

Manufacturer/Designer

Wrigley

The bottle pack is one example of the innovative packaging that Wrigley uses.

The bottle is recyclable, and last year, the bottle was re-engineered to become even lighter.

An innovative approach incorporating talc as a 'filler' decreased the amount of plastic used by 50 percent and the overall bottle weight by over 25 percent, while still maintaining the integrity of the package.

The UK has embraced the bottle pack by incorporating it into the Extra® portfolio and it is a popular format for Wrigley

FMCG INNOVATIONS

GREEN PACKAGING

Light Cap

FMCG INNOVATIONS

94

Manufacturer/Designer

Tetra Pak Ltd.
Bedwell Road,
Cross Lanes LL13 0UT,
Wrexham
Phone: +44 1978 834 000
Fax: +44 1978 834 001

LightCap 30; a high density polyethylene (HDPE) cap made from sugar cane.

Tetra Pak's TBA Edge, which is made from about 75% renewable resources, now comes with bio-based cap,

The renewable polyethylene used in LightCap 30 starts out as sugar cane.

The cane is crushed and the juice fermented and distilled to produce ethanol.

Through a process of dehydration, ethanol is converted into ethylene, which is then polymerised to produce the polyethylene used to manufacture

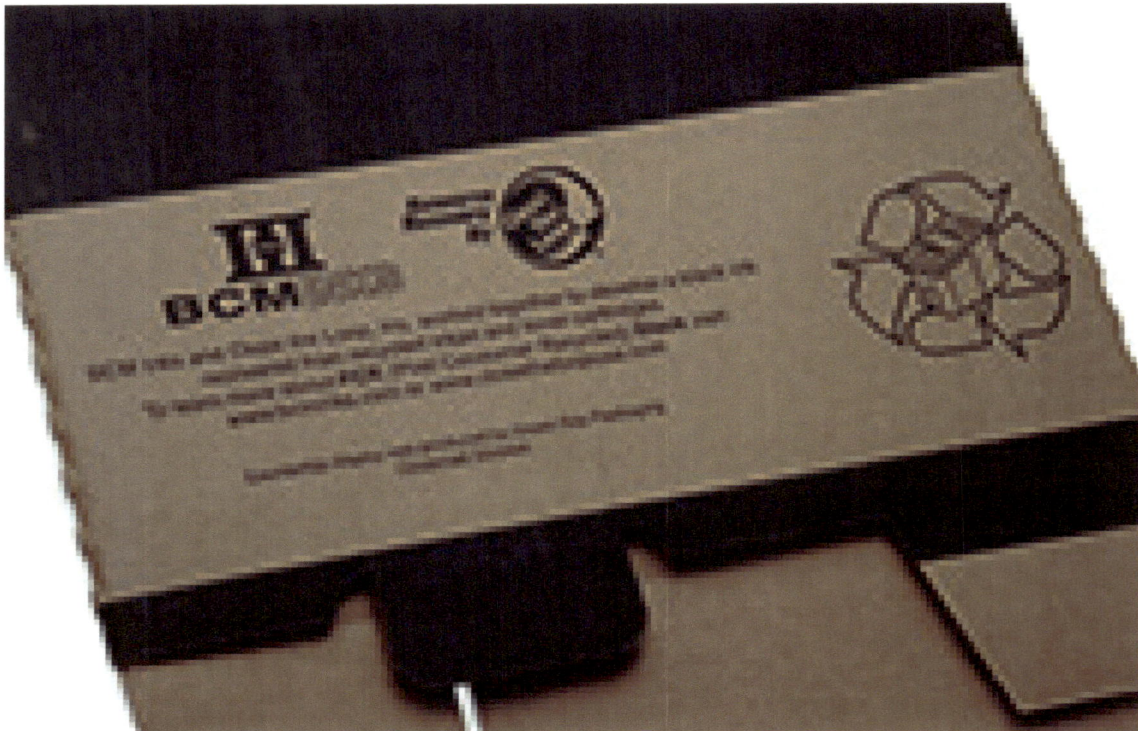

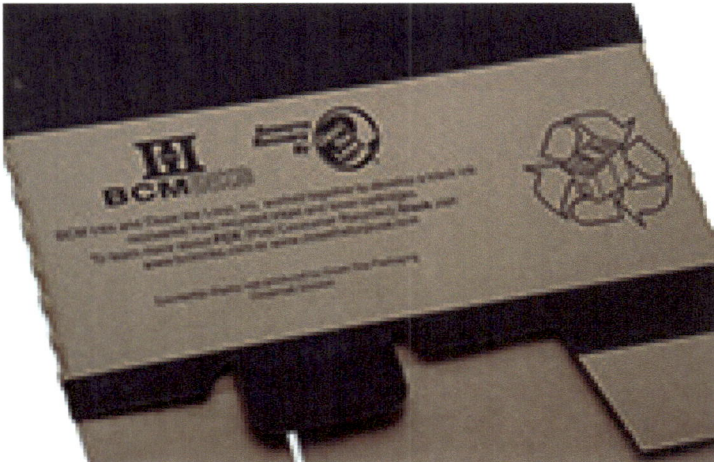

Manufacturer/Designer

BCM Inks, USA; Close the Loop. Ltd., USA
11400 Deerfield Rd
Cincinnati, OH 4 5242-2107
(513) 469-0400 Cincinnati office

"Ink Producer, Recovery Company Collaborate on Zero Waste to Landfill Program "

BCM Inks USA, Inc. partnered with Close the Loop materials recovery company to create a closed-loop system for used inkjet cartridges that result in zero waste to landfills.

Close the Loop recovers the plastic and remainder ink from used cartridges.

BCM Inks uses the recovered ink to develop a new water-based flexographic ink called 'PCR (Post-Consumer Recycled) Black' to print on corrugated shipping containers.

One drum of PCR Black stops approximately 200,000 ink cartridges from going into a landfill.

This recycling process recovers about 30 million ink cartridges a year."

Manufacturer/Designer

Sprint
http://www.sprint.com

Recyclable, boxes are made from unbleached kraft paper, using a minimum of 30% post-consumer recycled material.

Packaging is printed with soy inks and uses eco-friendly adhesives and aqueous coatings.

Addressing concerns from petroleum-based inks to printed user guides (which are now available online),

Packaging has successfully addressed an area that, until now, has largely been ignored within the telecommunications industry.

According to the study, for each million devices produced, the greening of Sprint packaging currently saves:

- The ecosystem equivalent to about two football fields of clear-cut forest

- 2,100 metric tons of carbon dioxide— the amount emitted by 420 passenger cars annually

- 8,800 megawatt hours of energy— enough to light the Statue of Liberty and Ellis Island for 12 months

- 8,900 kilogallons of water—enough to fill 68 million half-liter plastic water bottles

Packaging is 60% smaller in volume and 50% lighter in weight., significantly reducing the number of plane flights and truck runs necessary to move them.

Bamboo replaces EPE cushioning

Manufacturer/Designer

**Lenovo , Beijing
China**

Lenovo's corrugated packaging is especially designed for its consumer notebook product line.

It is Lenovo's effort to introduce green mass production packaging to mainstream product lines.

On inside, a bamboo cushion is used to replace EPE. Bamboo is world's most sustainable resource , rapidly renewable, recyclable and biodegradable.

The internal structure has been redesigned for optimum usage of space, therefore reducing the size of the package and providing a benefit to logistics.

The new corrugated packaging is an invention from Lenovo that will be good

SELF READY

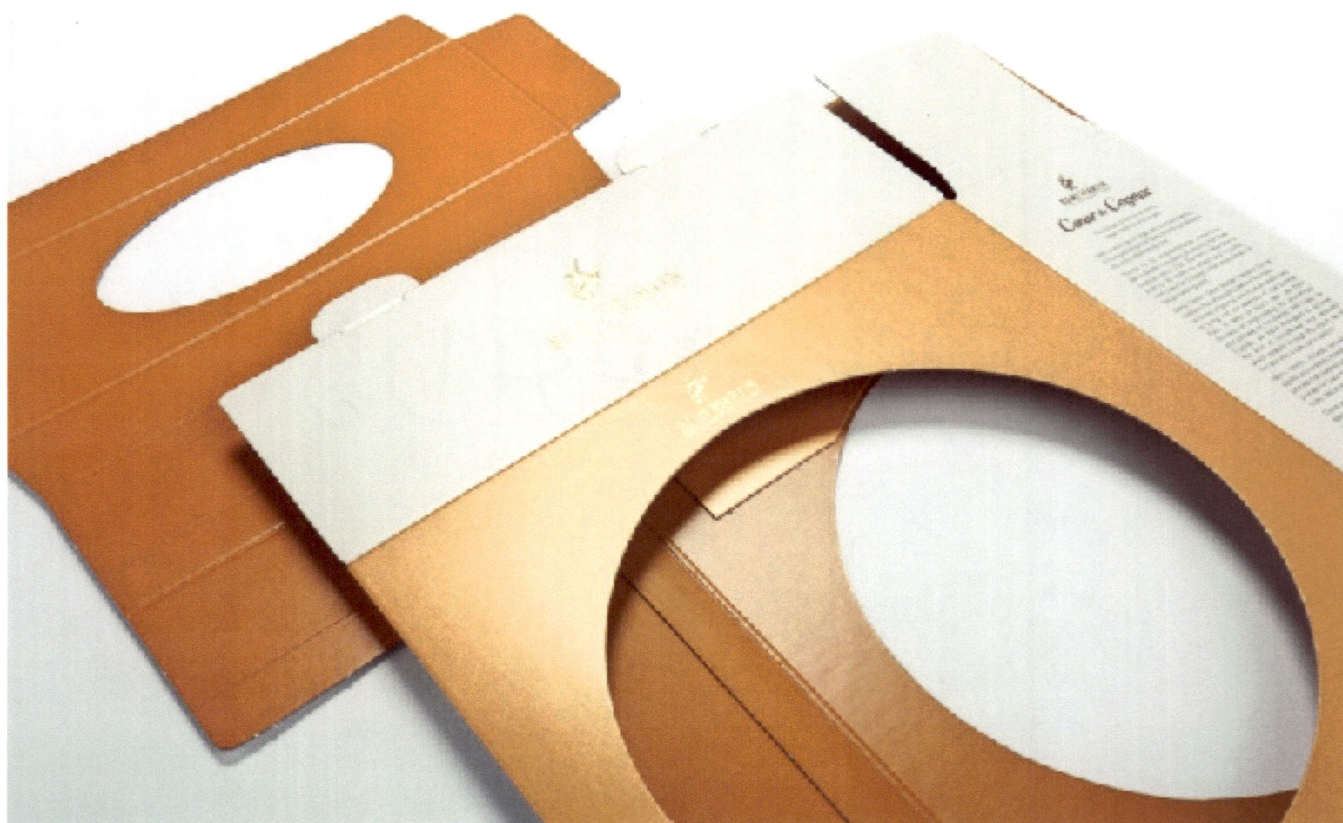

Manufacturer/Designer

Stora Enso Oyj Head Office
Kanavaranta 1
P.O.Box 309
FI-00101 HELSINKI
Finland
Tel. +358 20 46 131
Fax +358 20 46 214 71

Usually in cognac boxes the cartons are enclosed whereas on this one the bottle is displayed openly behind a large round cut out

A sustainability drive that led to a innovative shelf ready pack. A carton without plastic window.

"A folded box, open on both sides, which is designed to perfectly secure and protect the bottle, at the same time being clearly visible for consumers.

The mother of pearl gloss of the lacquer creates a soft glittering effect, giving the product added prominence.

Great visibility of the product allied to the safe and sturdy design ensured that this carton would appeal to consumers on the shelf.

FMCG INNOVATIONS

Manufacturer/Designer

Korsnäs AB
SE-801 81 Gavle
Sweden
Tel: +46 26 15 10 00
Fax: +46 26 15 22 40
Email: marketing@korsnas.com
URL: www.korsnas.com

A tidy stacking of ice cream bar.

Ice-cream bars could be placed in a folded carton and also that at least two cartons could be stacked next to each other. The basis for the packaging concept was based on a table display, a novel concept, as the selling of individual ice creams was not common practice in retailing, normally they are sold in multi-packs.

The new display is supplied closed to the trade. The upper part of the display can be removed via a tear strip and the eight ice cream bars are positioned neatly underneath.

carton offers a certain water-resistance and is also strong enough.

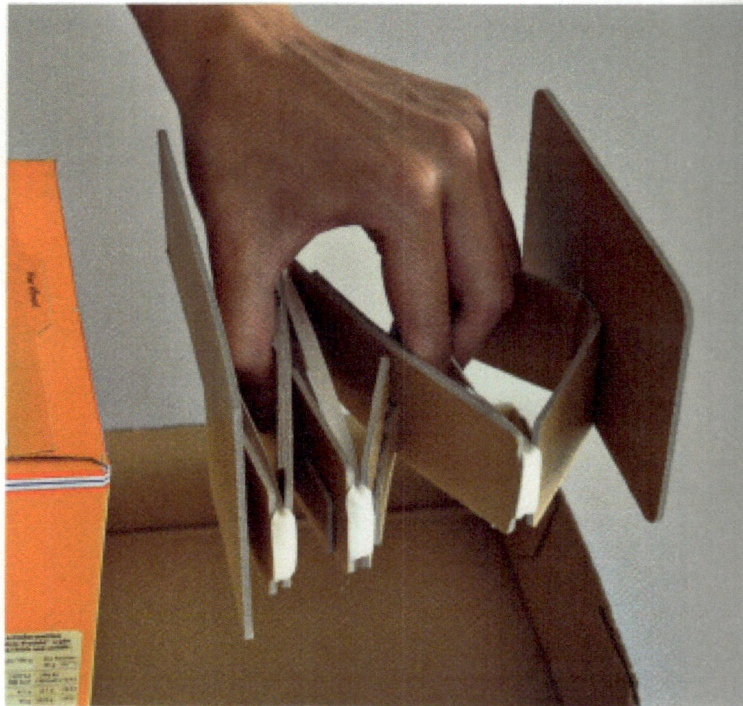

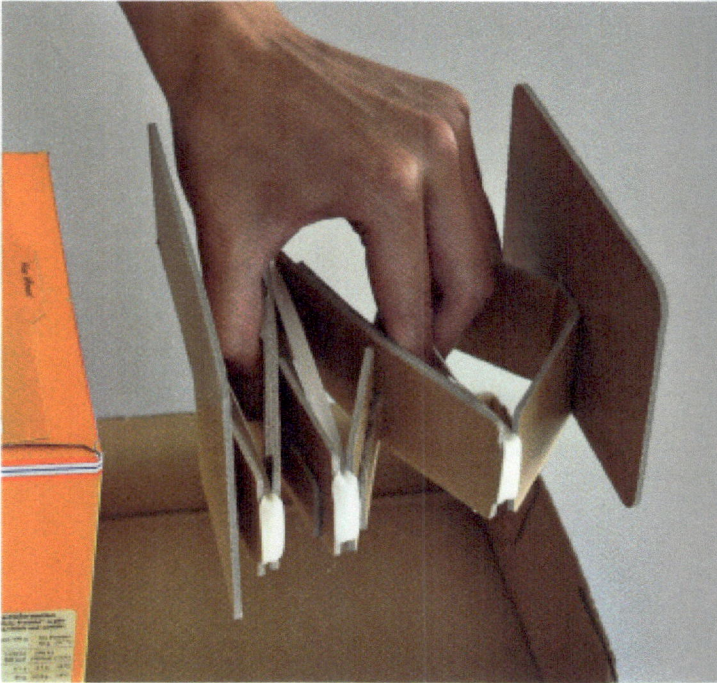

Manufacturer/Designer

THIMM Verpackung GmbH & Co. KG
Northeim, Germany

THIMM's POSe ® is a patented solution for a effective product presentation on the retailing shelf as well as in promotional displays.

When a product is removed the remaining items move forward automatically.

The system can flexibly be adjusted to different products and SRP's

Improves, Product recognition, better differentiation and less handling and is reusable.

Increases sales as well as shelf productivity at POS

FMCG INNOVATIONS

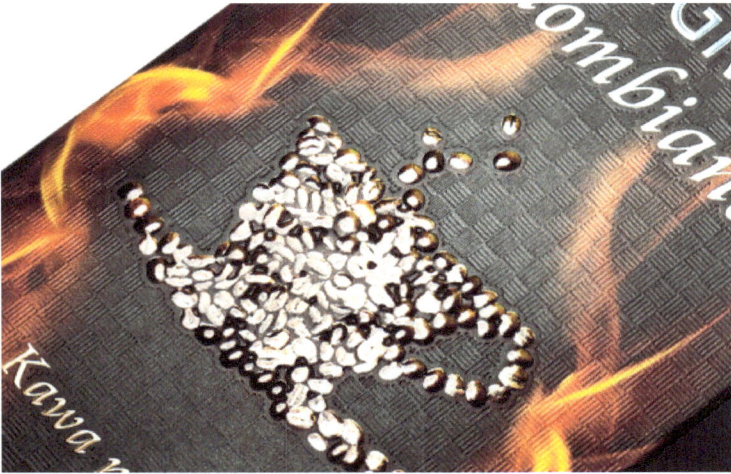

Manufacturer/Designer

International Paper Kwidzyn
Kwidzyn Mill
International Paper - Kwidzyn Sp. z
o.o.
Lotnicza 1
82-500 Kwidzyn
Poland
Tel.: +48 (0) 55 279 8000
Fax: +48 (0) 55 279 8451

Communicates the quality of the product and be as extraordinary as possible without requiring extravagant construction or unusual materials.

Using micro-embossing as a background, the design reminds of the textile structure typical of coffee sacks.

But to give that extra effect, they decided to employ a rarely used 3D tool: 3D-embossing of the beans.

Combination of 2D and 3D, finding both micro-embossing and the laminated carton with opaque white as well as matt and glossy varnish attractive.

A challenge in production was the depth of the 3D-embossing and the perfect register of all finishes and effects as the

Manufacturer/Designer

M-real Corporation
P.O. Box 224
Hallituskatu 1
33101 TAMPERE
Finland
Tel: +358 1046 337 71
Email: corporate.communica-
tions@m-real.com
URL: www.m-real.com

Unique elegance in the world of tea
Creating a bow with an angle of 10°
makes this box unique in the world of
tea.

The opening system is designed in the
shape of the Lipton logo.

The entire product range follows the
same graphic design with specific colours

Manufacturer/Designer

Scholz & Friends
Hamburg
Hanseatic Trade Center
Am Sandtorkai 76
20457 Hamburg
tel +49-40 3 76 81-0
Fax +49-40 3 76 81-681
info@sf.com

THE DIVERS WATCH IN WATER PACKAGING

The quality promised by Festina Profundo is that this watch stays waterproof. To visualise this claim this packaging design was developed

"We believe in what we see", the packaging aims to have maximum transparency.

The watch comes in a transparent bag filled with distilled water.

Featuring only the company logo and slogan, this packaging thus submits the product to quality control directly at the point of sale and serves as a mark of confidence at the same time.

By being subjected to this visible test of endurance, the watch is able to immediately convince the customers of its particularly high quality.

Manufacturer/Designer

Fanta, Australia

Thermochromic inks help launch Fanta's 'Unbottle the Fun' campaign -

Temperature-sensitive ink for promotion uses two levels of activation, for bottles both in chilled and ambient temperatures.

labels on Fanta Orange™ bottles were printed with thermochromic ink from Chromatic Technologies (CTI) that revealed a challenge to consumers to perform a "Funstigator" task,

Activation temperatures for the thermochromic inks based on whether the bottle was sold in the refrigerator section or the ambient shelf.

On the refrigerated bottles, the blue thermochromic inks would clear out as the beverage was consumed.

When the product is pulled from the refrigerator at 8°C (46°F) the ink is were fully colored, hiding the message behind.

On the flip side, the bottles sold at room temperature uitilises thermochromic ink set to be fully colored at 29°C (84°F) and relies on the heat of the consumers'

FMCG INNOVATIONS

Concept

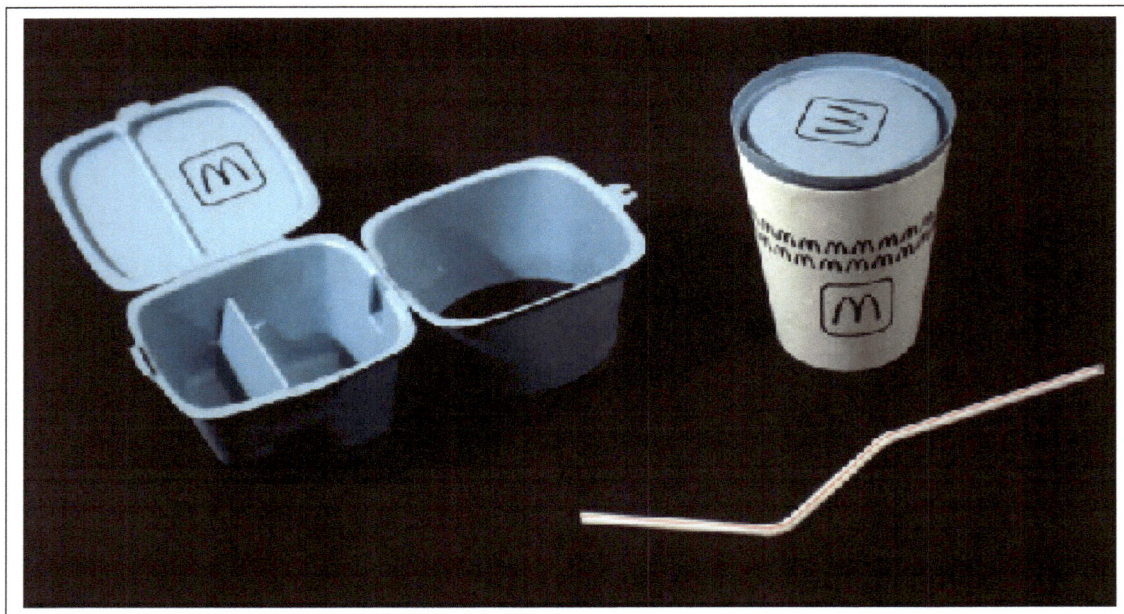

Manufacturer/Designer

Ian Gilley
The Elliott Group
Providence, RI, USA

Fabricated from biodegradable compressed paper.

Versatile system works on the go or at the table.

An attempt to reduce waste and make convenience food more convenient.

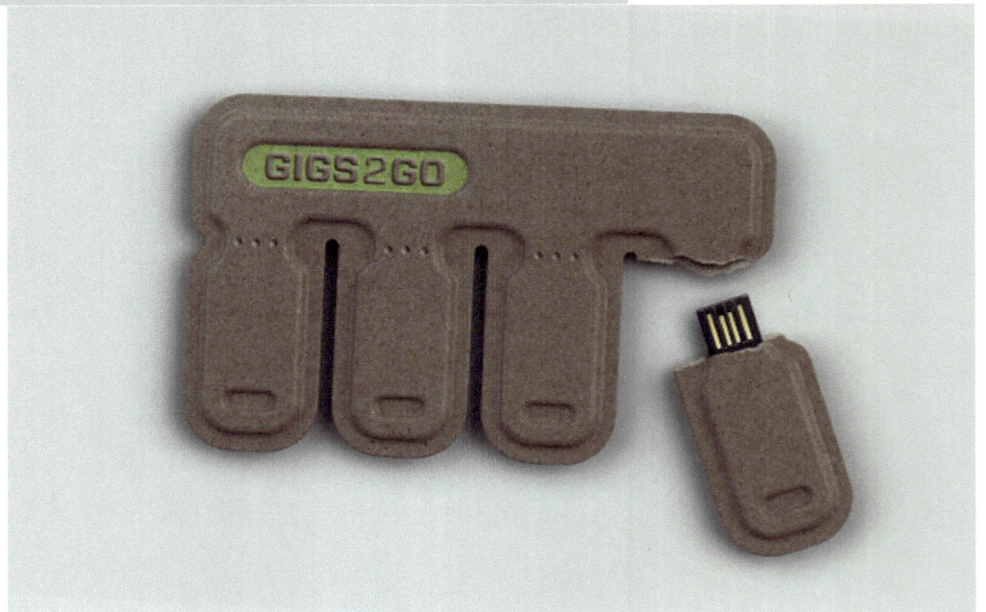

Manufacturer/Designer

BOLTgroup
GIGS.2.GO™

Made from molded paper pulp, this inexpensive, credit-card-sized data pack is a fast, easy way to share large files.

Portable "Tear-and-Share" Storage

When you need to share files on-the-go, simply tear off a tab from the GIGS.2.GO pack.

The molded paper pulp enclosure is made from 100% post-consumer recycled paper. It's renewable, biodegradable, lightweight, cheap, and durable enough to ensure that each tab

Manufacturer/Designer

Trican Co. Ltd
Hsinchu County, Taiwan

This design aims to create a tissue box that sits neatly and is accessible when at low capacity.

A flexible cardboard tissue tray has been especially designed to achieve this goal

The tray rises and falls from the Weight of the tissue

The design is cost effective.

Love Toothpaste tubes

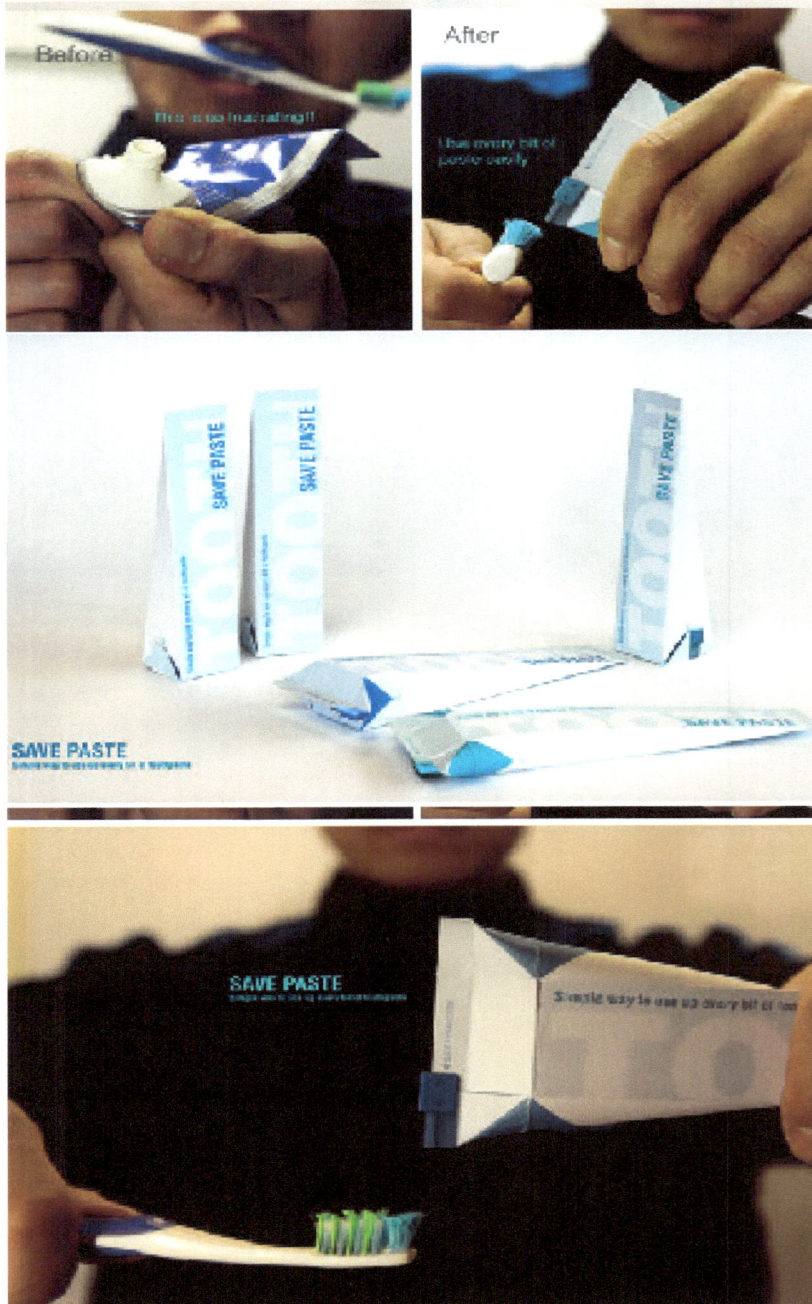

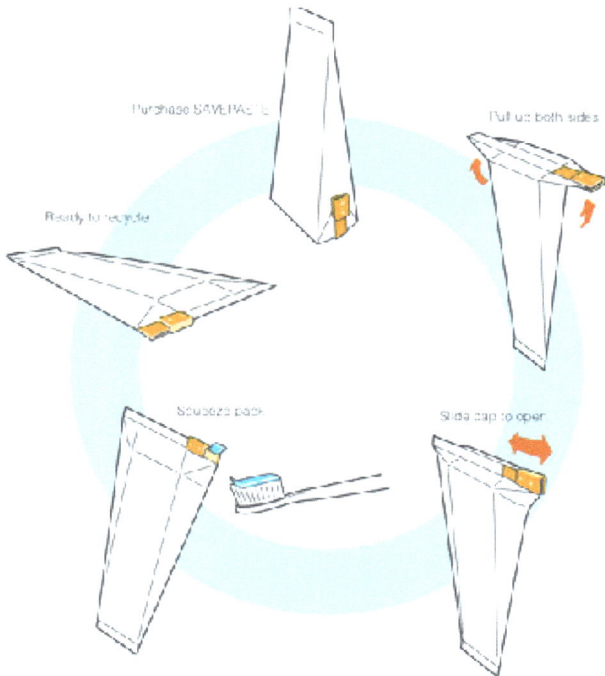

Purchase SAVEPASTE
Pull up both sides
Ready to recycle
Squeeze pack
Slide cap to open

Manufacturer/Designer

PERRY ROMANOWSKI

never liked toothpaste packaging.

Mostly because it never seems like you get all of the product out.

Then there are some people who don't squeeze from the bottom so the center gets all smashed and it looks awful.

Well, all that could end if companies started to adopt this innovative toothpaste packaging.

Before

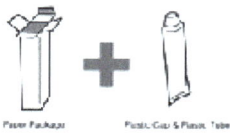

Paper Package

Plastic Cap & Plastic Tube

After

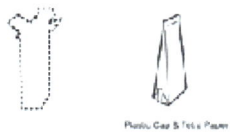

Plastic Cap & Foil Paper

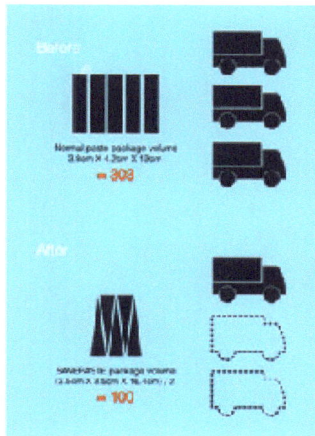

Comparing with traditional instant noodles bow ,Accordion Package saves more room,and the advantages can be seen in many ways:

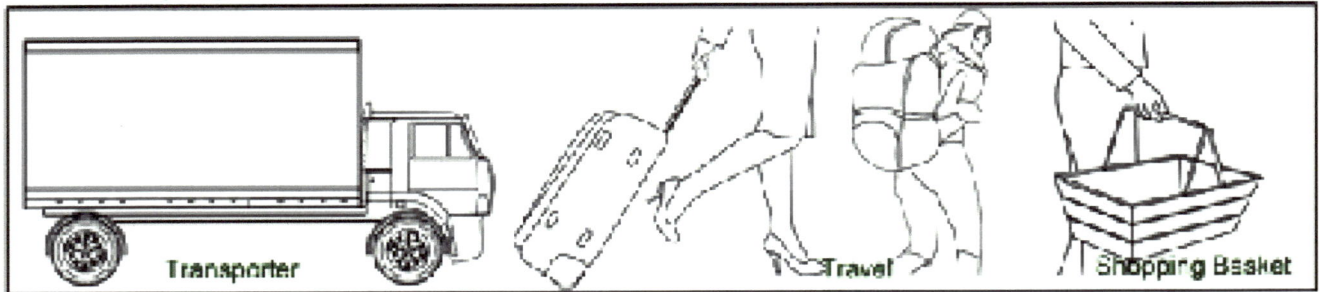

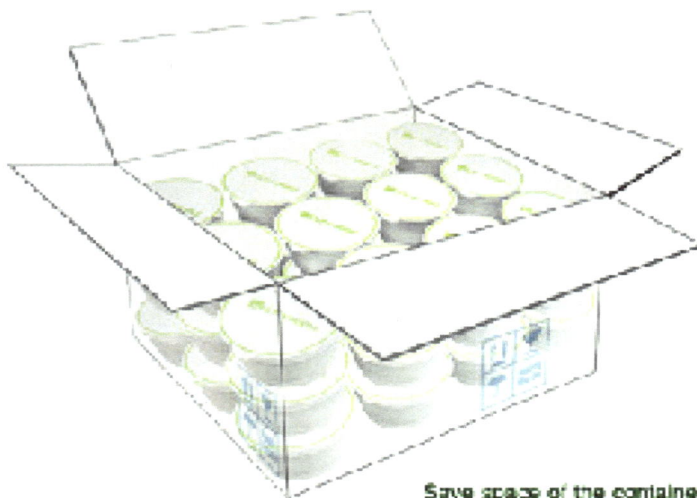

Transporter   Travel   Shopping Basket

Save space of the containers and reduce transportation costs.

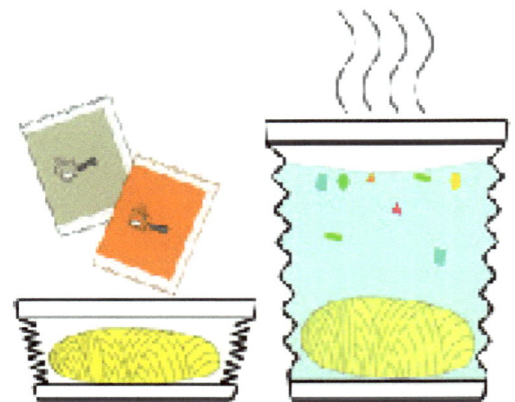

Accordion Package

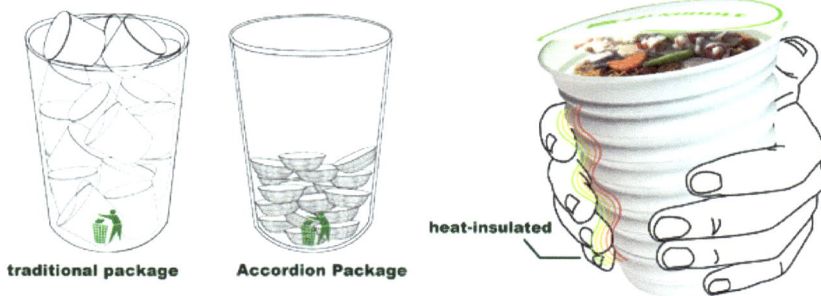

traditional package   Accordion Package

heat-insulated

Manufacturer/Designer

Liu Yi, Jiang Yuning & Luo Jing, South Korean

A collapsible noodle packaging Instant noodles and pastas are popular on-the-go products.

Traditionally packed in cup or bowl, taking up a lot of space in cupboards or in bags

The new container, baptized Accordion Noodle Package, for instant noodles, comes compressed into a small size with the dry noodles in it

While the consumer can stretch the accordion into a bigger cup to allow to add hot water and consume. After the noodles are eaten, the cup can be compressed to save space in the trash.

The accordion-like surface is said to decrease the contact points with the hands to prevent burning.

As the package has a smaller volume when it leaves the manufacturer's

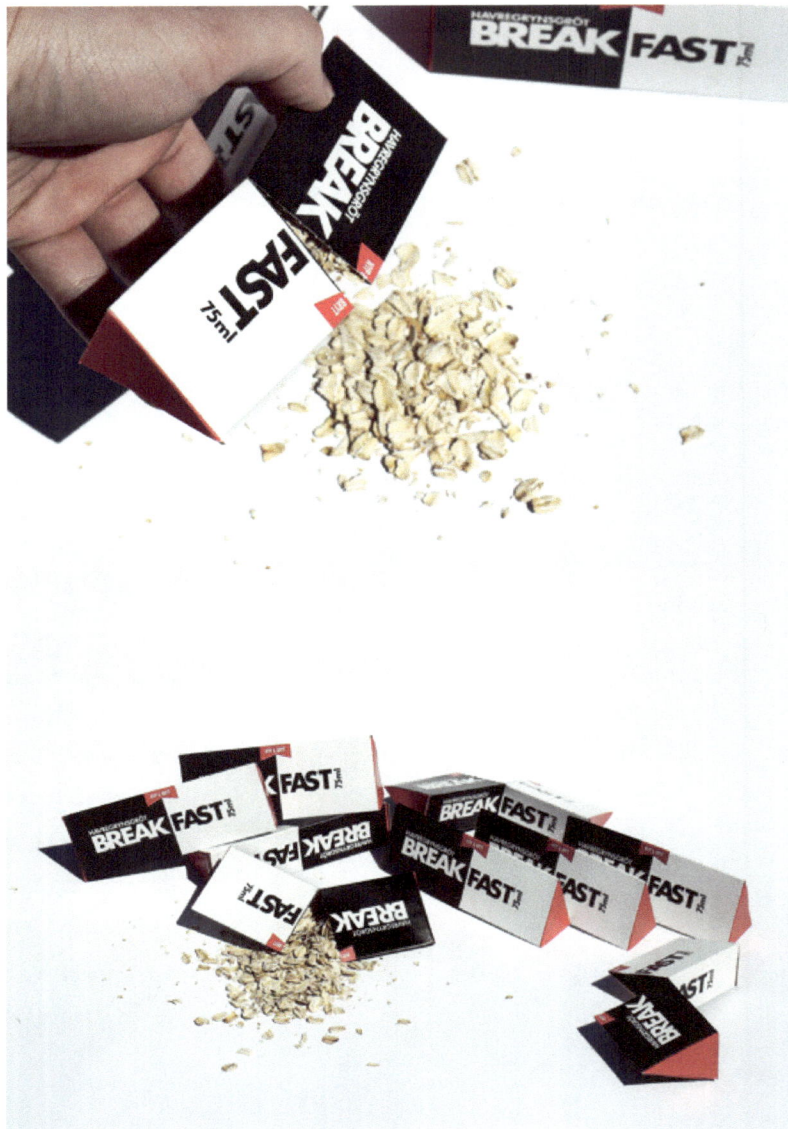

Manufacturer/Designer

Niklas Hessman, Sweden

A kind of taste pack for oatmeal.

"This package contains the right amount of oatmeal with added sugar and salt.

Break the BREAK FAST pack over a bowl, add water and cook in the microwave.

The idea is to target a new audience that otherwise would not eat oatmeal, or is in a hurry.

Break it fast and have a BREAK FAST!"

Manufacturer/Designer

Sta-Pack created by PT Bukit Muria Jaya/BMJ in
designed by Irvan Hermawan and M. Aidil
Saputra
Indonesia.

Sta-pack (Stacking Packaging)

The Sta-Pack consists of several separate units with an added handle.

A nice and elegant solution as fast food packaging, especially for the take-away market of upscale meals.

Stapack has unique characteristics. It adopted and modified the characteristics of the Tupperware elements and applied them to paperboard that can easily be folded saving storage space.

Each unit can hold a separate dish and only needs one handle for lifting as the wings of the individual boxes are closed until they locks. after which it is ready to be taken-out.

The Sta-Pack is made from food grade CDWB 230 gsm

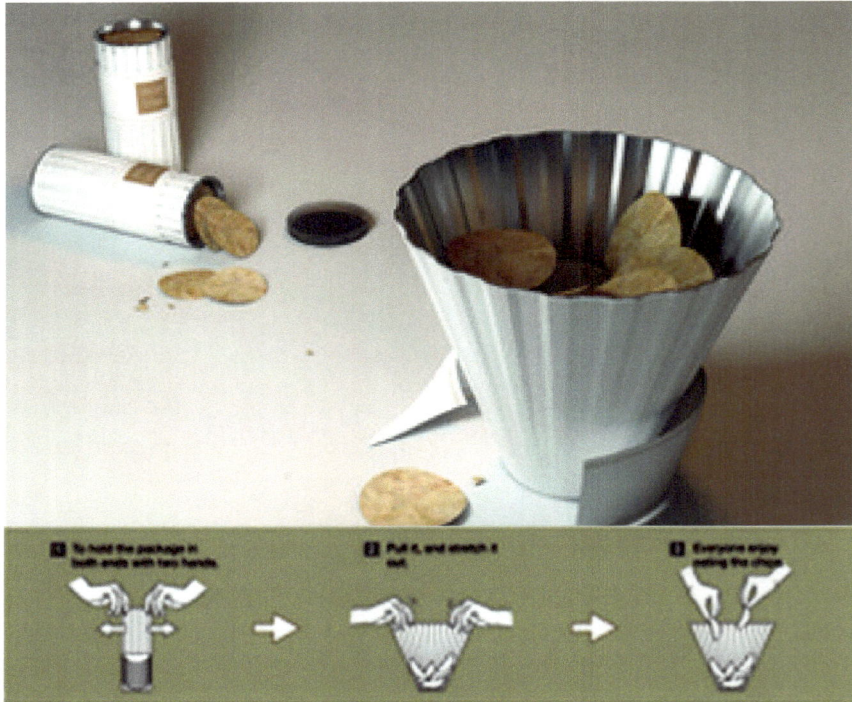

Manufacturer/Designer

designers Seok woo Kim, Dong han Seo, Do hyuk Kwon, Bum ho Lee, students at theKonkuk

The new innovation is called "Bloom Chips", as the package is blooming like a flower.

the moment the consumer opens the cylinder and folds out the sides like a blossom into a bowl-like shape.

While many a chips maker is anxious to enter the biodegradable pouch market, it should be better to invest in a bit more creativity.

Simply made from paperboard or an eco-friendly plastic, it is an improvement of the existing chips packaging (and certainly noiseless), more compact and with a better protection.

A party friendly option, can be used for serving and also for on the go consumptions.

FMCG INNOVATIONS

Thankyou

www.ingramcontent.com/pod-product-compliance
Lightning Source LLC
Chambersburg PA
CBHW041729210326
41598CB00008B/822